ENVIRONMENTAL CHEMISTRY

ENVIRONMENTAL CHEMISTRY

Arun Rastogi

ANMOL PUBLICATIONS PVT. LTD.
NEW DELHI-110 002 (INDIA)

ANMOL PUBLICATIONS PVT. LTD.
Regd. Office: 4360/4, Ansari Road, Daryaganj,
New Delhi-110002 (India)
Tel.: 23278000, 23261597, 23286875, 23255577
Fax: 91-11-23280289
Email: anmolpub@gmail.com
Visit us at: www.anmolpublications.com

Branch Office: No. 1015, Ist Main Road, BSK IIIrd Stage
IIIrd Phase, IIIrd Block, Bangalore-560 085 (India)
Tel.: 080-41723429 • Fax: 080-26723604
Email: anmolpublicationsbangalore@gmail.com

Environmental Chemistry

First Edition, 2010

ISBN 978-81-261-4512-6

PRINTED IN INDIA

Printed at Mehra Offset Press, Delhi.

Contents

Preface

The textbook of Environmental Chemistry is a dedicated book for all students, who are pursuing their Graduation or Post Graduation. The book covers the fundamentals of chemical and biochemical processes that take place in air, water, soil, and living systems. The text is written with advanced college students in mind; nevertheless, issues are frequently presented in such a way that graduate students and professionals can find subjects of interest.

Environmental Chemistry provides students with a thorough practical introduction to environmental concepts. The book emphasis on environmental literature by covering relevant data in many examples and problems.

Author

Chapter 1

Introduction

BACKGROUND

Environmental chemistry is the study of chemical processes occurring in the environment which are impacted by humankind's activities. These impacts may be felt on a local scale, through the presence of urban air pollutants or toxic substances arising from a chemical waste site, or on a global scale, through depletion of stratospheric ozone or global warming. The focus in our courses and research activities is upon developing a fundamental understanding of the nature of these chemical processes, so that humankind's activities can be accurately evaluated.

The field of environmental chemistry is both very broad and highly interdisciplinary. Within the Department of Chemistry we have a core group of faculty whose research interests are in atmospheric and aquatic chemistry, photochemistry, and the chemistry and transport of long-lived pollutants. We interact with other chemists in the Department, with numerous other researchers at the University who have related interests, and with nearby government agencies. Indeed, the setting for the study of environmental chemistry is ideal.

Our graduate programme consists of graduate courses which stress the fundamental photochemical, kinetic, analytical and transport aspects of environmental phenomena, regular seminars, and close interactions between the different research groups. We emphasize that students are able to put expertise in their own research field into a global context.

The field of environmental chemistry is rapidly expanding, and excellent employment opportunities exist in the academic, government, industrial and public policy sectors. The term environment refers to the natural elements that make up the Earth and surround living organisms, enabling life.

The study of the environment and its proper conservation, has gained enormous importance in last three decades. The term environmental pollution has become one of the most important issues of global concern.

ACCESSIBLE COMPONENTS OF ENVIRONMENT

- *Atmosphere*: This is a blanket of gaseous layer around the Earth.
- *Hydrosphere*: This includes all sources of water like oceans, rivers, lakes, glaciers, ground water, etc. and comprises about 96% of Earth's surface.
- *Lithosphere*: This comprises of soil, minerals, organic matter etc. and refers to Earth's solid crust containing the outer mineral cover.
- Biosphere: This is the domain of living organisms in contact with atmosphere, hydrosphere as well as lithosphere.

Man has begun to effect significant changes in his natural environment mainly after the advent of the Industrial revolution.

The change in industrial and agricultural practices in the last 300 years and the direction of overall economic, technological and institutional development, has led to a tremendous increase in the use of energy and its impact on the environment.

Due to these climatic changes, health hazards, effects on food and water supplies, changing habitats and eco-systems are effecting plants and animals life. Effects on the environment such as ozone depletion, greenhouse effect, acid rains, etc. have surfaced in recent years and are often the hot topics of discussion. The branch of chemistry, which deals with the study of various chemical phenomena occurring in the environment, is called environmental chemistry.

CONCEPT AND SCOPE

CONCEPT

Environmental chemistry" is the scientific study of the chemical and biochemical phenomena that occur in natural places. It should not be confused with green chemistry, which seeks to reduce potential pollution at its source. It can be defined as the study of the sources, reactions, transport, effects, and fates of chemical species in the air, soil, and water environments; and the effect of human activity on these. Environmental chemistry is an interdisciplinary science that includes atmospheric, aquatic and soil chemistry, as well as heavily relying on analytical chemistry and being related to environmental and other areas of science.

Environmental chemistry is that branch of one,which deals with the study of chemical and biochemical phenomena that occur in natural places like air,soil,water.

Environmental chemistry involves first understanding how the uncontaminated environment works, which chemicals in what concentrations are present naturally, and with what effects. Without this it would be impossible to accurately study the effects humans have on the environment through the release of chemicals.

Environmental chemists draw on a range of concepts from chemistry and various environmental sciences to assist in their study of what is happening to a chemical species in the environment. Important general concepts from chemistry include understanding chemical reactions and equations, solutions, units, sampling, and analytical techniques.

SCOPE

In chemistry the term mole represents a pile or mass of atoms, molecules, ions or electrons. Just as common man measures quantity in terms of kilograms or dozens, a chemical scientist deals with a 'mole' of atoms, molecules, ions or electrons. A mole is the amount of a substance that contains the same number of units of atoms or molecules or ions, as there are atoms in 12 g of the carbon,

$$^{12}_{6}C.$$

The 12 g of carbon contains a certain number of atoms of carbon. This number is estimated to be 6.023×10^{23} and is called Avogadro's number. Thus 12 g of the carbon,

$$^{12}_{6}C$$

is one mole of carbon.

Since the number of atoms in 12 g of carbon is called Avogadro's number, a mole may be redefined as the amount of a substance, which contains Avogadro number of molecules.

Avogadro's number is the number of elementary units (atoms, molecules, or ions) present in one mole of a substance, and it is equal to 6.023×10^{23}.

The concept of a mole being equal to a number,

6.023×10^{23}

is akin to the concept of a dozen being equal to a number 12.

A mole of atoms consists of Avogadro's number of atoms, and has a mass equal to one gram atomic mass of that element. *Therefore, the following illustrations hold true:*

- One mole of oxygen atoms contains 6.023×10^{23} atoms of oxygen and has a mass of 16 g.
- One mole of sodium atoms contains 6.023×10^{23} atoms of sodium, and has a mass of 23g.

Similarly one mole of molecule of a substance has Avogadro's number, 6.023×10^{23}, of molecules and has a mass equal to one gram molecular mass of that substance.

- One mole of oxygen molecules contains 6.023×10^{23} molecules of oxygen and has a mass of 32 g.
- One mole of sodium (mono atomic) contains 6.032×10^{23} molecules (or atoms, as it is mono atomic) and has a mass of 23 g.
- One mole of sulphuric acid contains 6.023×10^{23} molecules of sulphuric acid and has a mass of 98 g.
- One mole of carbon dioxide contains 6.023×10^{23} molecules of carbon dioxide, and has a mass of 44 g.
- One mole of an ionic substance like sodium chloride

or calcium chloride contains 6.023×10^{23} formula units.

- One mole of sodium chloride contains 6.023×10^{23} formula units i.e., 6.023×10^{23} sodium ions and 6.023×10^{23} chloride ions. It has a mass of 58.5 g (i.e., 23 + 35.5) g.
- One mole of calcium chloride contains 6.023×10^{23} formula units i.e., 6.023×10^{23} calcium ions and [2 (6.023×10^{23})] chloride ions. It has a mass of 111 g.
- One mole of aluminium chloride contains 6.023×10^{23} formula units of it, and contains 6.023×10^{23} aluminium ions and $(3 \times 6.023 \times 10^{23})$ chloride ions. It has a mass of 133.5 (27 + {3 × 35.5}).

Normally, mass of 1 mole of a substance is called its molar mass. It is expressed as g/mol. For example molar mass of water is equal to 18 g/mol.)

COMPONENTS OF ENVIRONMENT

Components of environment can be Analysed at various levels i.e. At the level of activities, at the level of processes, and at the level of orientations. Some of these issues and examples are discussed below. As discussed the following activities have been suggested as components of environmental quality.

- Fire protection.
- Comfort of home.
- Electric service.
- Privacy in your home.
- Relation with fellow workers.
- Postal service.
- Garbage-collection.
- Mechanical helpers in your family.
- Telephone services.
- Public-water system.
- Relation with Neighbours.
- Gas service.
- Freedom to live where you want.
- Sewage disposal service.
- Availability of food around your living place.

- Convenience for getting to important places.
- Noise level in the home setting.
- Beauty of your home.
- Security of your home.
- Topography of land around.
- Product available to the community.
- Medical care in your locality.
- Police protection.
- Quality of water used by the household.
- Natural outdoor recreation.
- Variety of wild life in community.
- Cleanliness of air around.
- Overall weather.
- Public information media.
- Crowding in your residential Neighbourhood.
- Transportation over long distance.
- Level of traffic congestion.
- Job-opportunities.
- Quality of water for recreation.
- Product quality and variety.
- Relations among group in the community.
- Freedom to move from class to class.
- Freedom to move form on job to another.
- Public services-gas, sewage etc.
- Unspoiled natures.
- Man made environment.
- Physical condition of environment at school.
- Isolation of your community.
- Physical condition of environment where you work.
- Amount of open space around.
- Access to parks.
- Control of dogs and other pets.
- Cost of living.

Understanding environment involves studying the complex relationships between the people and the typical physical settings in which they conduct their daily lives. Environmental interest in studying human behaviour in the familiar, everyday physical environment where people live and work as well as its relevance to the environment design and social planning has made it especially responsive to the demands of today's world.

Environmental studies are an area of social sciences where the focus of investigation is the interrelationship between the physical environment and human Behaviour and experience of man.

Fig below presents this.

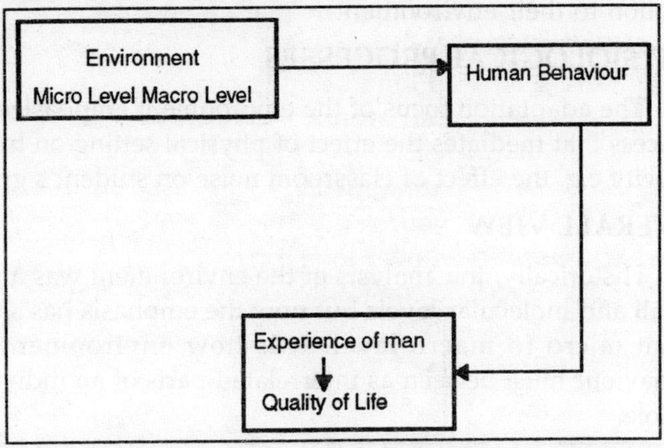

The other perspective of components of environment relates to environmental processes, its multidisciplinary and applied aspects.

ENVIRONMENTAL PROCESSES

Environment is new subject and a complex field of study. Therefore it is important to consider some of the characteristics that describe the what, how and why of the ways environmental processes work. In this chapter we attempt to explain the adaptation focus, physiological processes, the holistic view, interdisciplinary involvement and applied orientation of the subject matter.

Adaptation Focus

The process of adaptation. It is of interest to study how people adapt to the complex demands of the physical environment. For example, how people live in crowded places or overcrowded setting. Robert White defines adaptation as encompassing all the processes with their environment. This includes most simple ways of dealing with minor environment and irritations to the most complex efforts to cope with major environment changes. These challenges in living system in interaction with the environment are the adaptation processes. The holistic view of the organism and environment is considered along with the active role of living organism in relation to their environment.

PHYSIOLOGICAL PROCESSES

The adaptation focus of the environment emphasizes the process that mediates the effect of physical setting on human activity e.g. the effect of classroom noise on student's grades.

OVERALL VIEW

Historically, the analysis of the environment was at very small and molecular levels but now the emphasis has shifted from micro to macro level. It is how environment and Behaviour must be seen as interrelated parts of an indivisible whole.

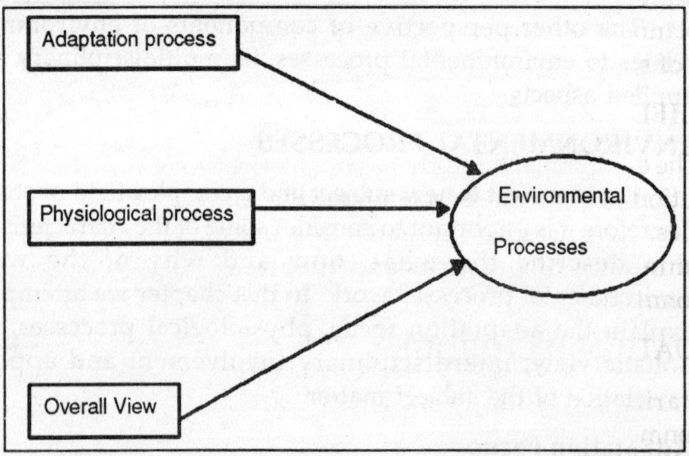

The positive and adaptive ways in which people cope with environmental challenges suggest active roles. This view looks into how people have varied and creative ways to cope with their environments.

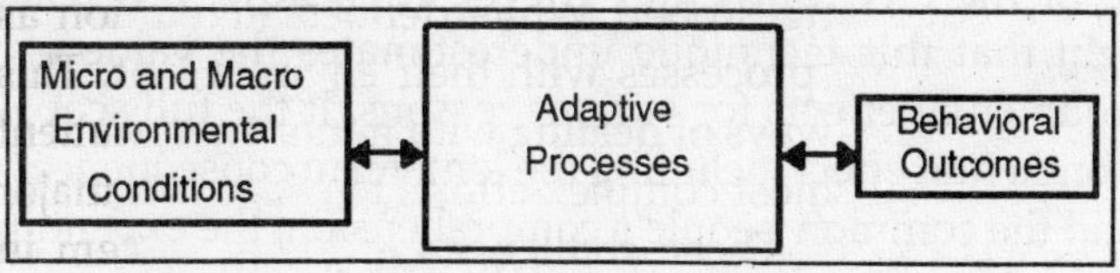

The adaptation process model (shown above) shows how environment effect on Behaviour is mediated by a number of processes.. It shows that the direction of effect in the environment-Behaviour relationship is reciprocal i.e. people may act on the environmental conditions while the environment in turn also acts on human Behaviour. The negative effects of situation such as over- crowding may be reversed through effective coping processes.

MULTIDISCIPLINARY ORIENTATION

The branches of study here have included not only environment but also other fields such as Sociology, Psychology, Geography, Anthropology, Medicine, Architecture and Planning etc (see figure above). The study of human Behaviour in physical settings requires the works of researchers in many social sciences as well as of architects and planners responsible for the design of human settings. The areas of study include.

APPLIED ORIENTATION

The environmental study has orientation towards both the resolution of practical problem and the formulation of new theories. Lewins action research is a useful model for the intermingling of practical and theoretical needs of the environment.

VALUATION TECHNIQUES

Various techniques, which attempt to means environmental effects, are below.

Preventive Expenditure Techniques

How much people are prepared to spend to prevent degradation of environment can be used as a measure of the value of the environmental effects. It is, however, usually thought that this technique underestimates the value of the environmental effects for various reasons. If the full scopes of environmental effect including the long-team consequences are beyond the common people's comprehension, the expenditure they would undertake to prevent degradation of environment would all very much short of the required levels. Secondly, it may always not be possible to fully offset all environmental effects by preventive expenditure especially in case of massive investment projects (such as large dams) which cause large environmental changes.

Replacement Cost Techniques

In this technique, how much people are prepared to spend to restore the environment to its previous state after degradation has occurred is taken as a measure of the value of the environmental effects. This technique also suffers from shortcomings similar to those mentioned in the context of the preventive expenditure technique. It is normally beyond the competence of common people to judge the full scope of damages done to the environment and it is not usually possible to restore the environment to its previous state after it has been damaged.

Property Value Techniques

Real estate values depend on various factors, including environmental factors. A residential accommodation at the lake or with a view to maintain commands a premium in the market. Air pollution depresses price of property in the area. People are willing to pay more for property similar in other respects in areas with clean air.

The difference in market price can be used as a measure of the value to society of the cleaner environment. The measurement of this value in practice may however be quite difficult.

In order to separate the effect of cleaner environment on properly value from other factors (such as location, quality of neighbourhood, etc) that affect a site's value, one would require to collect data as the value of various sites throughout the area and analyse the data with the help of an appropriates statistical method.

There have been several studies exploring the relationship between air pollution and property values, most of these studies show a significant negative relationship between properly value and air pollution. These studies have, however, been criiicized for underestimating the environmental effects of air pollution. People often do not know or understand the severity of pollution problems at different sites.

It requires special expertise to accurately estimate the pollution level at a site. Even if people are told about the presence of air pollutants in the area, they cannot make any use of such information because they lack knowledge about the effects of pollutants on human health or property. In such conditions it is not logical to expect that difference in properly value would correctly estimate the value of environmental effects of air pollution.

Human Health - Effect Technique

One method to estimate the value of environmental effects as to study the effect of pollutant exposures on human health.

It involves putting a monetary value on illness and death caused by exposure to pollutants. This idea of putting money value on human life does not appeal to everyone. Many regard human life to be of infinite value. But, is society's behaviour always consistent with such thinking? IS society prepared to pay whatever is necessary to prevent a death? If human life were considered of infinite value, we would never allow those activities which pose a risk to human beings, given that virtually no economic activity is completely risk-free, and they all would be stopped.

For example, injurious and deaths caused by road accidents could be reduced significantly if laws pertaining to road safely are made very strict. Several studies have shown that appropriate changes in design of vehicles and lowering of speed limits can reduce the number and severity of traffic accidents significantly. But we know that society does not take all these required steps. This ofcourse, does not mean that human life is cheap. Most people consider it invaluable, if not of infinite value.

One way to determine the effects of exposure to pollutants on human health is to study the incidence of diseases among people. Of course, air pollution is not the role cause of diseases. There are many other explanatory variables such as age, sex, hygienic conditions, climate, living standard and dietary habits. In order to separate the effect of air pollution from other variables, a large body of data on the incidence of various diseases among different sections of the population in various places needs to be collected and analysed with the help of appropriate statistical methods.

An alternative method of determining the effects of exposure of pollutants is to conduct Labouratory experiments on plants or animals. Plants animals are exposed to different levels and different kinds of pollutants to study their effects. Rats are often used as study objects for various reasons. The results of such studies, however, cannot be generalized too much. These studies cannot tell much about the effects of pollutants on human health as the human physiology is much different from that of rat or any other animal.

Once the relationship between pollutant exposures and human health effects is determined, the next step is to estimate the value of human health effects in monetary terms. The human health effect technique calculates the value of the health effects by looking at lost economic production and increased health-care costs.

The opportunity costs of an individual's premature death can be calculated by estimating the discounted present value of his future earning if he had not died prematurely. This provides a measure of lost economic production and thus a measure of loss to society.

In case of illness, the cost of care is added to the losses of earnings. This technique however does not value the psychic costs of ill health.

Many people find such procedure of calculating the worth of human life highly unsatisfactorily because according to this criterion a large section of people would command a very low economic value, especially the old people, disabled and those unemployed or in poorly paid jobs.

Travel Cost Method

This method estimates the time and money that people are willing to spend on 'eco-trips', i.e., visits to places of environmental value such as mountains, forests, lakes and parks. This method, however, severely underestimates the worth of the environmental assets and services because of its highly narrow focus.

It only looks at the entertainment, recreational aspects of environmental assets. Environment constitutes the basic life support system; its total value consists of several components as discussed above. This approach ignores most of them.

Contingent Valuation

In this method, the value of environmental assets and services is assessed through public opinion surveys. People are directly asked their valuation of the environment. They are asked how much thing is willing to pay to conserve the environments.

There are problems, however, with approach. People often lack knowledge in that area; therefore their responses do not accurately reflect their concerns. Secondly, there is a general inherent problem with Survey method i.e. the answer to a question very much depends on the way it is asked. If people realise that their answer will have effect on the decisions they may reply in strategic manner.

For example, if how much they are willing to pay is a hypothetical question (i.e., no payments are involved), people will answer it in one way. On the other hand, if there is a probability that they may actually have to pay they will answer it in another way. In the later case, there would be a systematic underestimation of the value of the environment. If the understand that they do not have to pay, the polluter will pay, their responses are likely to overestimate the value of the environment. If they are asked how much compensation they should be given for the environmental damages, they would certainly exaggerate the worth of the environment.

For these reasons, it is quite difficult to interpret the survey responses and come up with a figure which genuinely represents people's valuation of the worth of the environment. Investigators should try to frame questions in such a manner which induces individuals to respond not as interested parties but as disinterested observers, as people concerned with the public interest. For example The study of university housing environment is one research area where theoretical and practical objectives have been successfully combined. Thus one may understand the components of environment at various levels i.e. At the level of activities, at the level of processes and at the level of orientations.

ENVIRONMENT & ECOLOGY

A Systems Perspective

In the earlier sections it has been explained that there are many components of the concept and these concerns have always been there ever since the pre-historic time. The fact is that environment and ecology do not stand in isolation. Man. Society and culture are affected and in turn affect these concerns. These could be viewed as the systems analysis. A systems thinking suggest that all subsystems of the main system interact with each other and if one sub system doesn't make impact it affects the total system. The systems analysis could further be studied as a closed system analysis and an open system analysis. Fig below presents a closed system analysis.

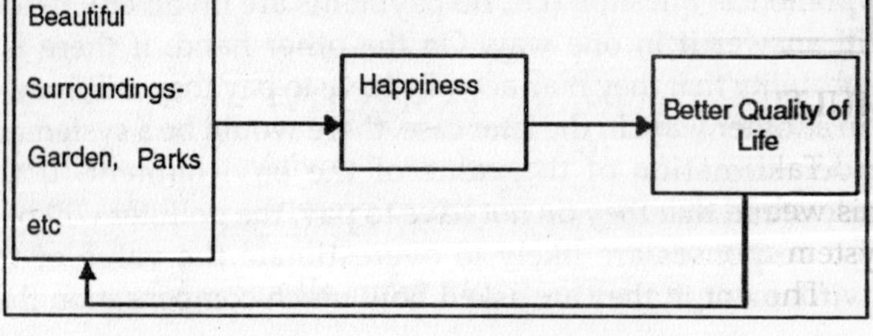

Similarly, there are ongoing impacts of environment on the quality of life of people, when time moves on. In fact, systems approach, views interactions and interdependence among subsystems are important as individual components. If one of the subsystems malfunctions, it affects the performance of the entire system.

Boulding (1956) has pointed out different levels of systems ranging from static structures to cybernetic structures and human systems.

He believes that human systems and social systems are most complicated ones and we have yet to discover the rudiments of theoretical models for them. Keeping these analogies and components analysis in mind one could further analyse ecology and environment within a systems perspective.

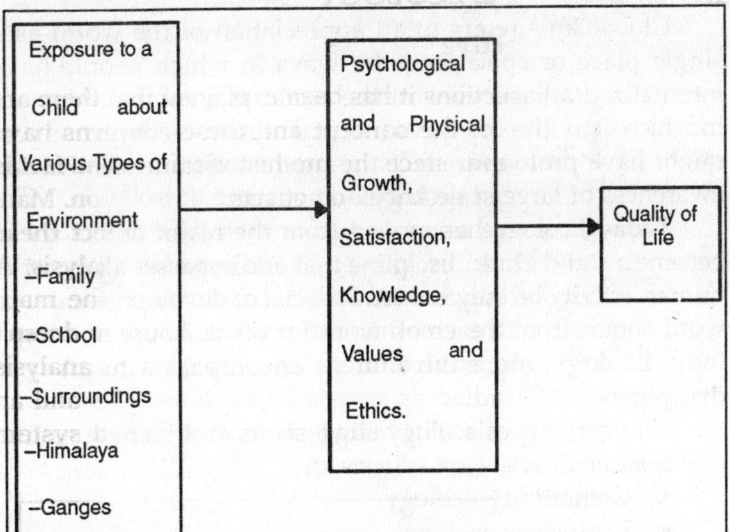

MULTIDISCIPLINARY PERSPECTIVE

Taking these as components of environment and ecology this web based course attempts to present a multidisciplinary system approach.

The concepts are drawn from all subjects relevant to

explain the environment scenario. Appropriate Technology, Psychology, Sociology, Economics, Management, and Philosophy, Values and Ethics are some subjects from where concepts and concerns are drawn.

The course also presents methodological approaches, training, ideologies, globalization, ecological behaviour and knowledge management relating them to environment and ecology.

The 21st century is a century of shorter distances, closer places but one that is plagued by a multitude of problems, man made as well as natural disasters, catalyzed by man made actions.

The growing populace with its increased technology use is one that is seemingly empowered to control its actions and yet susceptible to the minutest happenings as were its forefathers.

Globalism - refers to an appreciation of the world as a single place or space and the ways in which people have internalized a basic view of the world as a common domain in which activities or events (natural or man-made)in one part might have profound effects in another, distant one. It is an awareness of large scale knock-on effects.

Today Ecology has moved from the realm of biology to become a stand alone discipline that encompasses all facets of human activity be they economic social or developmental. The word comes from the greek work for oikos, house and logi - logy. Ecology and environment encompass a number of disciplines.

An overview of ecology suggests its multivariate nature. *Some of the ecological systems are*:

- Community ecology
- Ecosystem ecology
- Biological ecology & diversity
- Geographical ecology
- Oceanographic & marine ecology
- Ecosystems of the world
- Evolutionary ecology
- Political ecology

ENVIRONMENT: A MULTIDISCIPLINARY PERSPECTIVE

Environment as a concept is this multidisciplinary involving a number of disciplines as presented earlier in fig which is repeated here.

Social Science	Technology
Economic	FactorsPolicy related factors
Social factors	Psychological factors

Environmental analysis also includes the understanding of multiple issues related to disciplines.

Some of the examples are:

- Air pollution
- Water pollution
- Industrial pollution
- Social environment
- Psychological environment
- Biological environment
- Ozone depletion
- Global climate change
- Rise in sea level & other natural changes
- Antarctica: a source for resource development
- Marine resources
- Offshore oil & gas
- Manmade habitats etc

This calls for:

- Studies of sustainable development
- Environmental politics
- Environmental audit
- Suitable environmental policies & legislation Therefore, in this section we have visualized an overall perspective of ecology and environment from multidisciplinary angle. The next section presents some of the methodological considerations to Analyse these issues.

ENVIRONMENT &ECOLOGY ANALYSIS

Scientific Social Research

It would be difficult to comprehend the nature and content of research without an appreciation of the method we designate as scientific.

Research, simply put is an endeavour to discover inteilectual and practical answer to problems through the application of scientific methods to knowable universe. The scientific methods increase the likelihood of answering questions related to WHAT, HOW And Why in an enquiry.

Einstein and Infield state the essence of science in the following words - "Science is the attempt of the human mind to find a connection between the world of ideas and world of phenomena.

All the essential ideas in science were born in dramatic conflict between reality and our attempt at understanding the same." This does not mean, however, that science will always come out with an answer..... Thompson observes "The belief that science has explained everything is hopeless misunderstanding." There may be many paths leading to truth, however scientific is entails more understandable and reliable. Meaning of Research therefore is 'A careful, critical inquiry or examination in seeking facts or principles in order to as certain something.

Scientific Social Research

It would be difficult to comprehend the nature and content of research without appreciation of the methods we designate as scientific.

According to Scientific Social Research have the following properties: Reliance on empirical evidence Use of relevant concepts Commitment to objectivity Ethical neutrality Predictions based on probability Testing of conclusions through replication.

Research Process

There are various measures to study the effects of,

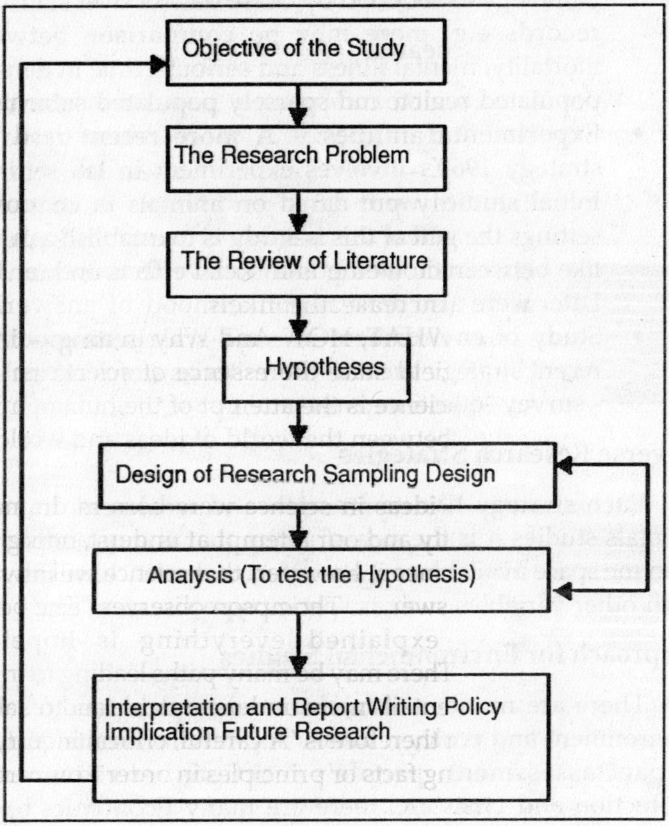

environment. Earlier strategies were not that sophisticated and reliable, thus mere have been fascinating development in this field.

- Co relational studies.
- Experimental studies in Labouratory settings.
- Small scale fields settings.
- Diverse research strategies.
- Case studies.
- Evaluation studies.
- Econometric studies.
- Time series studies.
- Correlation studies based census data: It is based on study relationship between density and social

pathology as reflected by cause data or over statistical records e.g. mere may be comparison between mortality, mental illness and serious crime in densely populated region and sparsely populated suburbs.

- Experimental studies: - A more recent research strategy 1960's involves experiment in lab settings. Initial studies were dared on animals in controlled settings the god of this is study is to establish a casual like between crowding and social effects on humans. Later were also done as humans.
- Study of environment in small-scale setting: - Most recent strategy such as - dormitories stores etc include - survey -questionnaires etc.

Diverse Research Strategies

Each strategy has its strength and weakness. In some animals studies it is the size of group that is more important man me space available to it also effect of density have inserted with other variables such as species type situation etc.

Approach for Environmental Studies

There are number of models and approaches to study environment and ecology. Measuring Environmental Values, impact assessment, people's reactions, environmental protection and crisis etc. there are many Economics based approaches issues and approaches.

They are:

- Market values
- Indirect methods
- Contingent valuation

Some of the indirect methods are:

- Preventive expenditure
- Replacement cost technique
- Human capital technique
- Travel cost method Other Methods include
- Technical coefficient and the value of Human Life
- Epidemiological studies

- Property value and Air pollutants
- Survey techniques

Environmental issues are too vital to be left out of economic decision making. Social cost benefit analysis cannot be complete without taking into consideration environmental considerations. In this chapter we discuss various methods that have been suggested for measuring environmental values.

It is not easy to asses the environmental effects of a project. What types of physical environmental effects would a project cause and to what extent, can rarely be answered satisfactorily. There is always an element of uncertainty associated with there effects. Most large investment projects now require environmental impact assessment (EIAs) that estimates the environmental effects of the projects. Often these EIAs churn out estimates of the effects, which are not only incomplete but also, devoid of scientist rigor.

Even if the physical effects are assessed with reasonable degree of accuracy and certainty, the next problem one encounters is how to assign economic values to these effects. It is a widely held view that a market economy does not value all aspects of the environment appropriately. Markets and prices may not exist for many of the environmental effects that require valuation.

Even if market prices exist, they may be distorted for various reasons. The environmental effects of big projects are very large.

In such cases, the project may actually alter prices. For these various reasons indirect methods are used for valuation of the environmental effects. There are a number of techniques developed by economists who attempt to value the environmental effects.

Before discussing these various techniques it needs to be noted that total economic value (TEV) of environmental assets consists of use and no-use values.

The total economic value has four components:

TEV = Direct use values + Indirect use values + Option value + Existence value

Let us consider the case of tropical forest. Direct values

are the resources and services provided directly by the fcrest such as habitat and food for indigenous people, timber and minor non-timber products. Indirect use values of environmental assets are essentially their ecological functions, which indirectly support economic activity and human welfare.

Tropical forests, for example, protect adjacent eco-systems, provide watershed protection, and are store of CO2. An option value is the value placed on an option that allows for use of the environment in the future.

The amount individuals would be willing to pay to conserve a tropical forest, or at least some of its uses, for future would be its option value.

Existence value of an environmental asset in the value placed on it independent of its any actual or possible future use. It relates to valuation of the resources as a unique asset in itself, an object of intrinsic value. Existence value of tropical forests arises because people simply want them to exist, even if they are never likely to ever see or use then in any way.

Environmental Impact Assessment: includes Environmental Inventory; Environmental Assessment; Environment Impact Statement; Historical Perspective; Environmental Assessment Process; Goals of EIA Process need to emphasize methodological issues related to following aspects:

- Air Environment.
- Water Environment.
- Noise Environment.
- Socio-Economic Environment.
- Cultural Environment.
- Biological Environment.
- People's Participation.
- Case Studies are also important in this regard. Some examples.
- Air Pollution Monitoring.

Ganga Action Plan

The study of Ganga: The Holy River; Ganga; From Gomukh to Bay of Bengal; Ganga; Its importance, Ganga; Overuse,

Abuse and Pollution; What is River Pollution? What are the Principal Sources of Pollution? How to measure the extent of pollution? Concern for the River; Formulation of the Ganga Action Plan; what does the Action Plan seek to do? River Front Facilities; Low Cost Sanitation; Construction of Electric Crematoria, Pollution from Industries; Scientific Content in the Context of Ganga Action Plan;Public Involvement in the Ganga Action Plan; Monitoring of Water Quality; Recycling Sewage; Where we went wrong. What do we need to do? Can people's participation help in this project?

Industrial Pollution: Some Case Studies: Pollution Control Aspects of Fertilizer Industries; Ammonia Plant Effluents; Removal of Carbon; Oil Removal; Hydrogen Sulphide Removal; Nitrogen-Wash Tail Gas; Ammonium Sulphate Plant; Phosphoric Acid Plant; Solid Effluent; Liquid Effluent; Gaseous Effluent; Pollution Control in Petroleum Refineries and Petrochemical Units; Characteristics of Liquid Effluent; Refineries; Petrochemical Industries; Refinery Liquid Waste-Treatment Methods; Implant Measures; Steam Stripping; Oxidation; Removal of Phenols; Biological Treatment; Disposal of Sludge's; Treatment of Liquid Effluents from Petrochemical Industries.

Environmental Audit: Case studies Indian Scenario Advantage of Audit; Environmental Audit.

How to measure the extent of pollution? Concern for the River; Formulation of the Ganga Action Plan; what does the Action Plan seek to do? River Front Facilities; Low Cost Sanitation; Construction of Electric Crematoria, Pollution from Industries; Scientific Content in the Context of Ganga Action Plan;Public Involvement in the Ganga Action Plan; Monitoring of Water Quality; Recycling Sewage; Where we went wrong. What do we need to do? Can people's participation help in this project?

Industrial Pollution: Some Case Studies: Pollution Control Aspects of Fertilizer Industries; Ammonia Plant Effluents; Removal of Carbon; Oil Removal; Hydrogen Sulphide Removal; Nitrogen-Wash Tail Gas; Ammonium Sulphate Plant; Phosphoric Acid Plant; Solid Effluent; Liquid Effluent;

Gaseous Effluent; Pollution Control in Petroleum Refineries and Petrochemical Units; Characteristics of Liquid Effluent; Refineries; Petrochemical Industries; Refinery Liquid Waste-Treatment Methods; Implant Measures; Steam Stripping; Oxidation; Removal of Phenols; Biological Treatment; Disposal of Sludge's; Treatment of Liquid Effluents from Petrochemical Industries.

CYCLES OF MATTER IN ENVIRONMENT

Environmental problems caused by human interference in matter cycles. In the past decades humans have been the cause of extensive environmental pollution.

Pollution is not a one-sided term; all kinds of matter has been applied by humans, resulting in a disturbance in natural processes.

Disturbances of natural processes are clearly shown, when you take a look at human interference in matter cycles. This has caused various environmental problems, which are important issues today.

HYDROLOGICAL CYCLE H_2O

There are three ways in which humans cause environmental problems by interfering with the hydrological cycle. Firstly, we remove large quantities of freshwater from rivers, lakes and groundwater supplies.

In areas where great amounts of people use the water from groundwater supplies, groundwater may be fully depleted. This effect is enhanced by vegetation removal to create new farmland or underground mines, or to build roads.

Secondly, because of vegetation removal, rainwater washes away and it no longer infiltrates, so that the groundwater supplies are not restored. In many cases groundwater supplies will than be filled up with salt water from rivers and lakes.

When groundwater is salinated the entire area is influenced. Other effects also occur because of vegetation removal; the risk of flooding is enhanced, and soil erosion and land sliding increase. Finally, people change the quality of water, by adding nutrients and contaminants. This causes the ecological processes that usually purify the water to be disturbed.

- *Precipitation*: Transported through the circulating atmosphere the clouds move themselves inland, as a result of gravity, and lose their water as it falls back unto the ground. This phenomenon is called rain or precipitation.
- *Infiltration*: Rainwater infiltrates into the ground and sinks to the saturated zone, where it becomes groundwater. Groundwater slowly moves from places of high elevation and pressure to places with low elevation and pressure. It moves from the area of infiltration through an aquifer and out to a discharge area, which can be either a sea or an ocean.
- *Transpiration*: Plants and other forms of vegetation take up water from the soil and excrete it again as water vapour. About 10% of the precipitation that falls on the ground vapourizes again through transpiration of plants, the rest evaporates from seas and oceans.
- *Surface run-off*: The rainwater that does not infiltrate into the soil will directly reach the surface water, as it will run-off to rivers and lakes. After that it will be transported back to the seas and oceans. This water is called surface run-off.
- *Evaporation*: Due to the influence of sunlight the water in oceans and lakes will warm up. As a result of that it will evaporate and rise up into the atmosphere. There it will form clouds that will eventually cause rainwater to fall back on earth. The evaporation of oceans is the most important kind of evaporation.
- *Condensation*: In contact with the atmosphere the water vapour will transform back to liquid, so that it will be visible in the air. These accumulations of water in the air are what we call clouds.

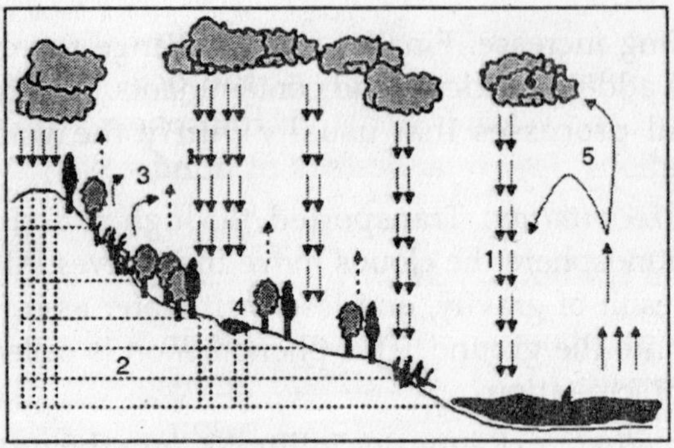

Fig. The Hydrological Cycle

Carbon Cycle

The geological component of the carbon cycle is where it interacts with the rock cycle in the processes of weathering and dissolution, precipitation ofminerals, burial and subduction, and volcanism (see our The Rock Cycle module for information).

In the atmosphere, carbonic acid forms by a reaction with atmospheric carbon dioxide (CO_2) and water. As this weakly acidic water reaches the earth as rain, it reacts with minerals at the earth's surface, slowly dissolving them into their component ions through the process of chemical weathering. These component ions are carried in surface waters like streams and rivers eventually to the ocean, where they precipitate out as minerals like calcite ($CaCO_3$). Through continued deposition and burial, this calcite sediment forms the rock called limestone.

This cycle continues as seafloor spreading pushes the seafloor under continental margins in the process of subduction. As seafloor carbon is pushed deeper into the earth by tectonic forces, it heats up, eventually melts, and can rise back up to the surface, where it is released as CO_2 and returned to the atmosphere.

This return to the atmosphere can occur violently through volcanic eruptions, or more gradually in seeps, vents, and CO_2-rich hotsprings. Tectonic uplift can also expose previously buried limestone. One example of this occurs in the Himalayas where some of the world's highest peaks are formed of material that was once at the bottom of the ocean. Weathering, subduction, and volcanism control atmospheric carbon dioxide concentrations over time periods of hundreds of millions of years.

Biological Carbon Cycle

Biology plays an important role in the movement of carbon between land, ocean, and atmosphere through the processes of photosynthesis and respiration. Virtually all multicellular life on Earth depends on the production of sugars from sunlight and carbon dioxide (photosynthesis) and the metabolic breakdown (respiration) of those sugars to produce the energy needed for movement, growth, and reproduction. Plants take in carbon dioxide (CO_2) from the atmosphere during photosynthesis, and release CO_2 back into the atmosphere during respiration through the following chemical reactions:

Respiration: $C_6H_{12}O_6$ (organic matter) + $6O_2$ $6CO_2$ + 6 H_2O + energy

Photosynthesis: Energy (sunlight) + $6CO_2$ + $H_2O \rightarrow C_6H_{12}O_6 + 6O_2$

Through photosynthesis, green plants use solar energy to turn atmospheric carbon dioxide into carbohydrates (sugars). Plants and animals use these carbohydrates (and other products derived from them) through a process called respiration, the reverse of photosynthesis. Respiration releases the energy contained in sugars for use in metabolism and changes carbohydrate "fuel" back into carbon dioxide, which is in turn released to back to the atmosphere. The amount of carbon taken up by photosynthesis and released back to the atmosphere by respiration each year is about 1,000 times greater than the amount of carbon that moves through the geological cycle on an annual basis.

On land, the major exchange of carbon with the atmosphere results from photosynthesis and respiration. During daytime in the growing season, leaves absorb sunlight and take up carbon dioxide from the atmosphere. At the same time plants, animals, and soil microbes consume the carbon in organic matter and return carbon dioxide to the atmosphere. Photosynthesis stops at night when the sun cannot provide the driving energy for the reaction, though respiration continues.

This kind of imbalance between these two processes is reflected in seasonal changes in the atmospheric CO_2 concentrations. During winter in the northern hemisphere, photosynthesis ceases when many plants lose their leaves, but respiration continues.

This condition leads to an increase in atmospheric CO_2 concentrations during the northern hemisphere winter. With the onset of spring, however, photosynthesis resumes and atmospheric CO_2concentrations are reduced.

This cycle is reflected in the monthly means (the light blue line) of atmospheric carbon dioxide concentrations shown in Figure 2.

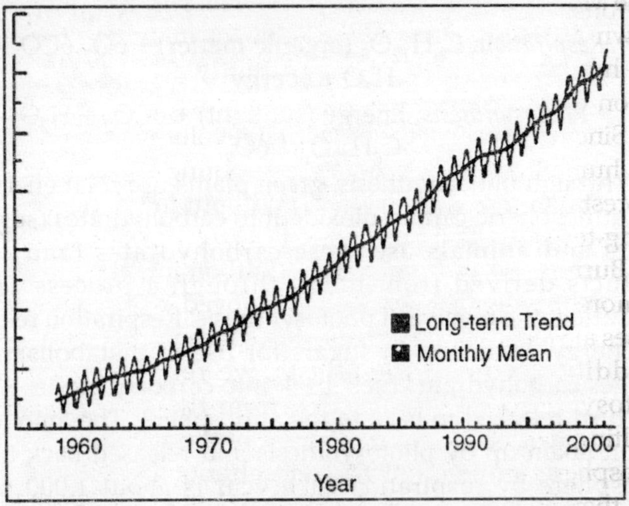

In the oceans, phytoplankton (microscopic marine plants that form the base of the marine food chain) use carbon to make shells of calcium carbonate ($CaCO_3$). The shells settle to the bottom of the ocean when phytoplankton die and are buried in the sediments.

The shells of phytoplankton and other creatures can

become compressed over time as they are buried and are often eventually transformed into limestone.

Additionally, under certain geological conditions, organic matter can be buried and over time form deposits of the carbon-containing fuels coal and oil. It is the non-calcium containing organic matter that is transformed into fossil fuel. Both limestone formation and fossil fuel formation are biologically controlled processes and represent long-term sinks for atmospheric CO_2.

ALTERATION OF CARBON CYCLE

Recently, scientists have studied both short- and long-term measurements of atmospheric CO_2 levels. Charles Keeling, an oceanographer at the Scripps Institute of Oceanography, is responsible for creating the longest continuous record of atmospheric CO_2 concentrations, taken at the Mauna Loa observatory in Hawaii. His data (now widely known as the "Keeling curve," shown in Figure 2) revealed that human activities are significantly altering the natural carbon cycle.

Since the onset of the industrial revolution about 150 years ago, human activities such as the burning of fossil fuels and deforestation have accelerated, and both have contributed to a long-term rise in atmospheric CO_2.

Burning oil and coal releases carbon into the atmosphere far more rapidly than it is being removed, and this imbalance causes atmospheric carbon dioxide concentrations to increase. In addition, by clearing forests, we reduce the ability of photosynthesis to remove CO_2 from the atmosphere, also resulting in a net increase. Because of these human activities, atmospheric carbon dioxide concentrations are higher today than they have been over the last half-million years or longer.

Because CO_2 increases the atmosphere's ability to hold heat, it has been called a "greenhouse gas." Scientists believe that the increase in CO_2 is already causing important changes in the global climate.

Many attribute the observed 0.6 degree C increase in global average temperature over the past century mainly to increases in atmospheric CO_2. Without substantive changes in global patterns of fossil fuel consumption and deforestation, warming trends are likely to continue.

The best scientific estimate is that global mean temperature will increase between 1.4 and 5.8 degrees C over the next century as a result of increases in atmospheric CO_2 and other greenhouse gases.

This kind of increase in global temperature would cause significant rise in average sea-level (0.09-0.88 meters), exposing low-lying coastal cities or cities located by tidal rivers such as New Orleans, Portland, Washington, and Philadelphia to increasingly frequent and severe floods.

Glacial retreat and species range shifts are also likely to result from global warming, and it remains to be seen whether relatively immobile species such as trees can shift their ranges fast enough to keep pace with warming.

Even without the changes in climate, however, increased concentrations of CO_2could have an important impact on patterns of plant growth worldwide.

Because some species of plants respond more Favourably to increases in CO_2 than others, scientists believe we may see pronounced shifts in plant species as a result of increasing atmospheric CO_2 concentrations, even without any change in temperature.

For example, under elevated CO_2 conditions, shrubs are thought to respond more Favourably than certain grass species due to their slightly different photosynthetic pathway. Because of this competitive inequality, some scientists have hypothesized that grasslands will be invaded by CO_2-responsive grass species or shrubby species as CO_2 increases.

In an attempt to understand whether recently observed changes in the global carbon cycle are a new phenomenon, or have instead occurred throughout geologic history, scientists have devoted considerable effort to developing methods for understanding Earth's past atmosphere and climate. These techniques include analysis of gas bubbles trapped in ice, tree rings, and ocean and lake floor sediments for clues about past climates and atmospheres.

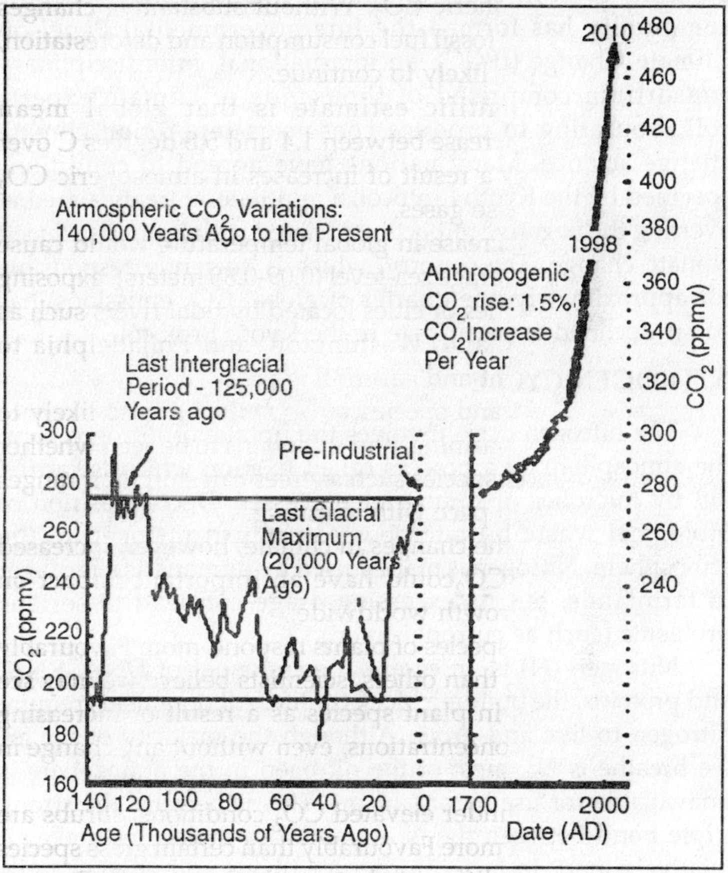

Together, these techniques suggest that over the past 20 million years, the Earth's climate has oscillated between relatively warm and relatively cold conditions called interglacial and glacial periods. During interglacial periods, atmospheric CO_2 concentrations were relatively high, and during glacial periods, CO_2 concentrations were relatively low. We are currently in an interglacial warm period, and human activities are pushing CO_2 concentrations higher than they have been for hundreds of thousands of years (Figure 3).

Understanding and mitigating negative impacts of atmospheric CO_2 enrichment constitute two of the most central challenges that environmental scientists and policy makers currently face. In order to address this issue, the scientific community has formed the Intergovernmental Panel on Climate Change (IPCC), an international, interdisciplinary consortium comprised of thousands of climate experts colLabourating to produce consensus reports on climate change science. Many nations have agreed to conditions specified by the Kyoto Protocol, a multilateral treaty aimed at averting the negative impacts associated with human-induced climate change. The country, which is currently responsible for approximately one quarter of global CO_2 emissions, has so far declined to participate in the Kyoto Protocol.

NITROGEN CYCLE

The nitrogen cycle involves the uptake of nitrogen form the atmosphere by a process called fixation which is carried out by microbes or industrial processes. Decomposition of biological waste by microbes can return nitrogen to the atmosphere. Nitrogen is mainly used by humans as a fertilizer in farmlands, but its excessive usage can lead to serious problems (such as eutrophication).

Nitrogen (N) is an essential component of DNA, RNA, and proteins, the building blocks of life. All organisms require nitrogen to live and grow. Although the majority of the air we breathe is N_2, most of the nitrogen in the atmosphere is unavailable for use by organisms. This is because the strong triple bond between the N atoms in N_2 molecules makes it relatively inert. In fact, in order for plants and animals to be able to use nitrogen, N_2 gas must first be converted to more a chemically available form such as ammonium (NH_4^+), nitrate (NO_3^-), or organic nitrogen (e.g. urea – $(NH_3)_2CO$). The inert nature of N_2 means that biologically available nitrogen is often in short supply in natural ecosystems, limiting plant growth and biomass accumulation.

Nitrogen is an incredibly versatile element, existing in both inorganic and organic forms as well as many different oxidation states. The movement of nitrogen between the atmosphere, biosphere, and geosphere in different forms is described by the nitrogen cycle (Figure below), one of the major biogeochemical cycles. Similar to the carbon cycle, the nitrogen cycle consists of various storage pools of nitrogen and processes by which the pools exchange nitrogen (arrows) (see our The Carbon Cycle module for more information).

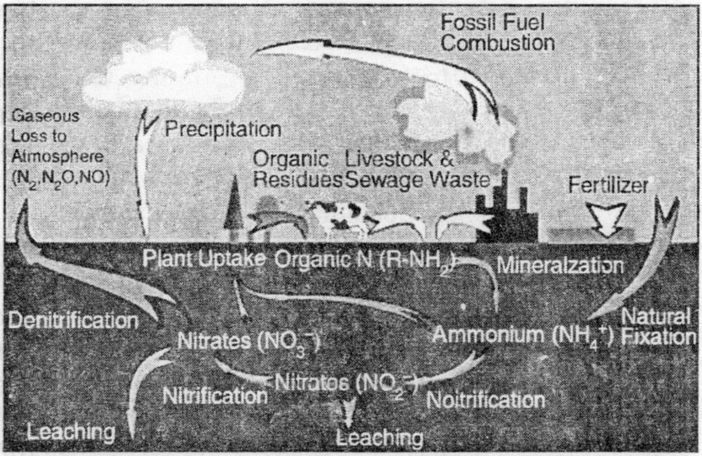

Five main processes cycle nitrogen through the biosphere, atmosphere, and geosphere: nitrogen fixation, nitrogen uptake (organismal growth), nitrogen mineralization (decay), nitrification, and denitrification. Microorganisms, particularly bacteria, play major roles in all of the principal nitrogen transformations. As microbially mediated processes, these nitrogen transformations tend to occur faster than geological processes like plate motion, a very slow, purely physical process that is a part of the carbon cycle. Instead, rates are affected by environmental factors that influence microbial activity, such as temperature, moisture, and resource availability.

Nitrogen Fixation

$N_2 \rightarrow NH_4+$

Nitrogen fixation is the process wherein N_2 is converted to ammonium, essential because it is the only way that organisms can attain nitrogen directly from the atmosphere. Certain bacteria, for example those among the genus Rhizobium, are the only organisms that fix nitrogen through metabolic processes. Nitrogen fixing bacteria often form symbiotic relationships with host plants. This symbiosis is well-known to occur in the legume family of plants (e.g. beans, peas, and clover). In this relationship, nitrogen fixing bacteria inhabit legume root nodules (Figure) and receive carbohydrates and a Favourable environment from their host plant in exchange for some of the nitrogen they fix. There are also nitrogen fixing bacteria that exist without plant hosts, known as free-living nitrogen fixers. In aquatic environments, blue-green algae (really a bacteria called cyanobacteria) is an important free-living nitrogen fixer.

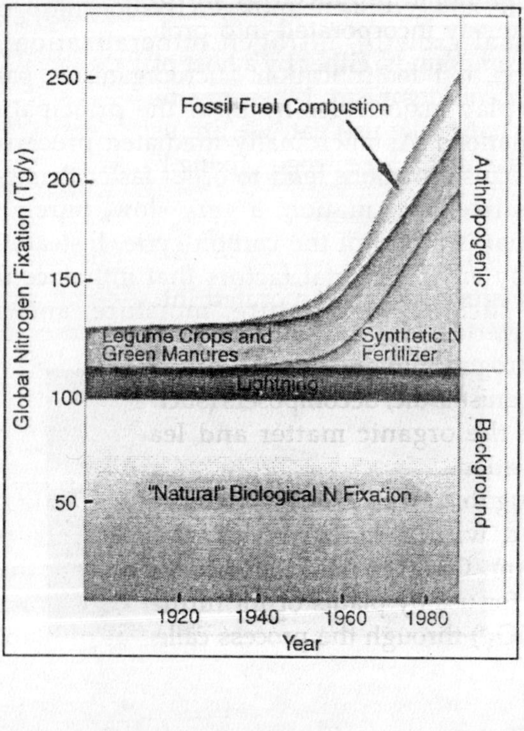

In addition to nitrogen fixing bacteria, high-energy natural events such as lightning, forest fires, and even hot lava flows can cause the fixation of smaller, but significant amounts of nitrogen (Figure above).

The high energy of these natural phenomena can break the triple bonds of N_2 molecules, thereby making individual N atoms available for chemical transformation.

Within the last century, humans have become as important a source of fixed nitrogen as all natural sources combined.

Burning fossil fuels, using synthetic nitrogen fertilizers, and cultivation of legumes all fix nitrogen. Through these activities, humans have more than doubled the amount of fixed nitrogen is pumped into the biosphere every year (Figure below), the consequences of which are discussed below.

Nitrogen Uptake

$NH_4+ \rightarrow$ Organic N

The ammonia produced by nitrogen fixing bacteria is usually quickly incorporated into protein and other organic nitrogen compounds, either by a host plant, the bacteria itself, or another soil organism. When organisms nearer the top of the food chain (like us!) eat, we are using nitrogen that has been fixed initially by nitrogen fixing bacteria.

Nitrogen Mineralization

Organic N $\rightarrow NH_4^+$

After nitrogen is incorporated into organic matter, it is often converted back into inorganic nitrogen by a process called nitrogen mineralization, otherwise known as decay. When organisms die, decomposers (such as bacteria and fungi) consume the organic matter and lead to the process of decomposition.

During this process, a significant amount of the nitrogen contained within the dead organism is converted to ammonium. Once in the form of ammonium, nitrogen is available for use by plants or for further transformation into nitrate (NO_3^-) through the process called nitrification.

Nitrification

$NH_4^+ \rightarrow NO_3^-$

Some of the ammonium produced by decomposition is converted to nitrate via a process called nitrification. The bacteria that carry out this reaction gain energy from it. Nitrification requires the presence of oxygen, so nitrification can happen only in oxygen-rich environments like circulating or flowing waters and the very surface layers of soils and sediments.

The process of nitrification has some important consequences. Ammonium ions are positively charged and therefore stick (are sorbed) to negatively charged clay particles and soil organic matter. The positive charge prevents ammonium nitrogen from being washed out of the soil (or leached) by rainfall. In contrast, the negatively charged nitrate ion is not held by soil particles and so can be washed down the soil profile, leading to decreased soil fertility and nitrate enrichment of downstream surface and groundwater's.

Denitrification

$NO_3^- \rightarrow N_2 + N_2O$

Through denitrification, oxidized forms of nitrogen such as nitrate and nitrite (NO_2^-) are converted to dinitrogen (N_2) and, to a lesser extent, nitrous oxide gas. Denitrification is an anaerobic process that is carried out by denitrifying bacteria, which convert nitrate to dinitrogen in the following sequence:

$NO_3^- \rightarrow NO_2^- \rightarrow NO \rightarrow N_2O \rightarrow N_2$.

Nitric oxide and nitrous oxide are both environmentally important gases. Nitric oxide (NO) contributes to smog, and nitrous oxide (N_2O) is an importantgreenhouse gas, thereby contributing to global climate change.

Once converted to dinitrogen, nitrogen is unlikely to be reconverted to a biologically available form because it is a gas and is rapidly lost to the atmosphere. Denitrification is the only nitrogen transformation that removes nitrogen from ecosystems (essentially irreversibly), and it roughly balances the amount of nitrogen fixed by the nitrogen fixers described above.

ALTERATION OF N CYCLE AND ITS CONSEQUENCES

Early in the 20th century, a German scientist named Fritz Haber figured out how to short circuit the nitrogen cycle by fixing nitrogen chemically at high temperatures and pressures, creating fertilizers that could be added directly to soil. This technology has spread rapidly over the past century, and, along with the advent of new crop varieties, the use of synthetic nitrogen fertilizers has led to an enormous boom in agricultural productivity.

This agricultural productivity has helped us to feed a rapidly growing world population, but the increase in nitrogen fixation has had some negative consequences as well. While the consequences are perhaps not as obvious as an increase in global temperatures or a hole in the ozone layer, they are just as serious and potentially harmful for humans and other organisms.

Not all of the nitrogen fertilizer applied to agricultural fields stays to nourish crops. Some is washed off of agricultural fields by rain or irrigation water, where it leaches into surface or ground water and can accumulate.

In groundwater that is used as a drinking water source, excess nitrogen can lead to cancer in humans and respiratory distress in infants. The U.S. Environmental Protection Agency has established a standard for nitrogen in drinking water of 10 mg per liter nitrate-N.

Unfortunately, many systems (particularly in agricultural areas) already exceed this level. By comparison, nitrate levels in waters that have not been altered by human activity are rarely greater than 1 mg/L.

In surface waters, added nitrogen can lead to nutrient over-enrichment, particularly in coastal waters receiving the inflow from polluted rivers. This nutrient over-enrichment, also called eutrophication, has been blamed for increased frequencies of coastal fish-kill events, increased frequencies of harmful algal blooms, and species shifts within coastal ecosystems.

Reactive nitrogen (like NO_3^- and NH_4^+) present in surface waters and soils, can also enter the atmosphere as the smog-component nitric oxide (NO) and thegreenhouse gas nitrous oxide (N_2O). Eventually, this atmospheric nitrogen can be blown into nitrogen-sensitive terrestrial environments, causing long-term changes.

For example, nitrogen oxides comprise a significant portion of the acidity in acid rain which has been blamed for forest death and decline in parts of Europe and the Northeast United States. Increases in atmospheric nitrogen deposition have also been blamed for more subtle shifts in dominant species and ecosystem function in some forest and grassland ecosystems.

For example, on nitrogen-poor serpentine soils of northern Californian grasslands, plant assemblages have historically been limited to native species that can survive without a lot of nitrogen.

There is now some evidence that elevated levels of atmospheric N input from nearby industrial and agricultural development have paved the way for invasion by non-native plants. As noted earlier, NO is also a major factor in the formation of smog, which is known to cause respiratory illnesses like asthma in both children and adults.

Currently, much research is devoted to understanding the effects of nitrogen enrichment in the air, groundwater, and surface water. Scientists are also exploring alternative agricultural practices that will sustain high productivity while decreasing the negative impacts caused by fertilizer use. These studies not only help us quantify how humans have altered the natural world, but increase our understanding of the processes involved in the nitrogen cycle as a whole.

Nitrogen is a part of vital organic compounds in microrganisms, such as amino acids, proteins and DNA. The gaseous form of nitrogen (N_2), makes up 78% of the troposphere.

One might think this means we always have plenty of nitrogen available, but unfortunately it does not work that way. Nitrogen in the gaseous form cannot be absorbed and used as a nutrient by plants and animals; it must first be converted by nitrifying bacteria, so that it can enter food chains as a part of the nitrogen cycle.

During the conversion of nitrogen cyano bacteria will first convert nitrogen into ammonia and ammonium, during the nitrogen fixation process. Plants can use ammonia as a nitrogen source.

Nitrogen fixation is carried out according to the following reaction:

$N_2 + 3\,H_2 \rightarrow 2\,NH_3$

After ammonium fixation, the ammonia and ammonium that is formed will be transferred further, during the nitrification process. Aerobic bacteria use oxygen to convert these compounds.

Nitrosomonas bacteria first convert nitrogen gas to nitrite (NO_2^-) and subsequently nitrobacter convert nitrite to nitrate (NO_3^-), a plant nutrient.

Nitrification is carried out according to the following reactions:

$2\,NH_3 + 3O_2 \rightarrow 2\,NO_2 + 2\,H^+ + 2\,H_2O$

$2\,NO_2^- + O_2 \rightarrow 2\,NO_3^-$

Plants absorb ammonium and nitrate during the assimilation process, after which they are converted into nitrogen-containing organic molecules, such as amino acids and DNA. Animals cannot absorb nitrates directly. They receive their nutrient supplies by consuming plants or plant-consuming animals.

When nitrogen nutrients have served their purpose in plants and animals, specialized decomposing bacteria will start a process called ammonification, to convert them back into ammonia and water-soluble ammonium salts. After the nutrients are converted back into ammonia, anaerobic bacteria will convert them back into nitrogen gas, during a process called denitrification.

Denitrification is carried out according to the following reaction:

$NO_3^- + CH_2O + H^+ \rightarrow ½\,N_2O + CO_2 + 1½\,H_2O$

Finally, nitrogen is released into the atmosphere again. The whole process starts over after release.

A schematic representation of the nitrogen cycle is shown here:

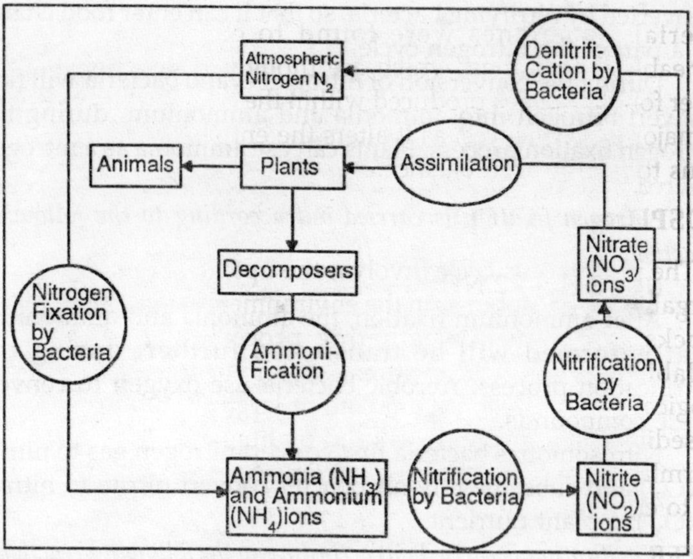

Nitrogen as a Limiting Factor

Although the nitrogen conversion processes often occurs and large quantities of plant nutrients are produced, nitrogen is often a limiting factor for plant growth. Water flowing across the soil causes this error. Nitrogen nutrients are water-soluble and as a result they are easily drained away, so that they are no longer available for plants.

Annamox Reaction

In 1999 researchers at the Gist-Brocades in Delft, The Netherlands, discovered a new reaction to be added to the nitrogen cycle; the so-called *annamox reaction*. This is now found to occur in the Black Sea, as well. The reaction implies conversion of nitrite and ammonium to pure nitrogen gas (N_2), which than escapes to the atmosphere. The reaction mechanism is triggered by a newly discovered bacterium, called *Brocadia anammoxidans*. This appears to be a compartmentalized bacterium; within the cell membrane two compartments can be found which are also surrounded by a membrane, a very rare phenomenon. Intermediate products of the reaction included hydroxylamine, and toxic hydrazine compounds. The bacterial membranes were found to consists of badly permeable membranes, which are thought to function as a barrier for hydrazines produced within the cell. This discovery has major consequences, as it alters the entire contribution of oceans to the nitrogen balance.

PHOSPHORUS CYCLE

The phosphorus cycle involves the uptake of phosphorus by organisms. Phosphorus in the environment is mainly found in rocks, and natural weathering processes can make it available to biological systems. After decomposition of biological waste, it can accumulate in large amounts in soils and sediments. Phosphorus is used by humans as a fertilizer in farmlands and in detergents. Overuse of phosphorus can lead to eutrophication.

WATER CYCLE

The water cycle is the process by which water travels in a sequence from the air (condensation) to the earth (precipitation) and returns to the atmosphere (evaporation). It is also referred to as the hydrologic cycle.

PHOSPHORUS CYCLE

Phosphorus is an essential nutrient for plants and animals in the form of ions PO_4^{3-} and HPO_4^{2-}. It is a part of DNA-molecules, of molecules that store energy (ATP and ADP) and of fats of cell membranes. Phosphorus is also a building block of certain parts of the human and animal body, such as the bones and teeth.

Phosphorus can be found on earth in water, soil and sediments. Unlike the compounds of other matter cycles phosphorus cannot be found in air in the gaseous state. This is because phosphorus is usually liquid at normal temperatures and pressures. It is mainly cycling through water, soil and sediments. In the atmosphere phosphorus can mainly be found as very small dust particles. Phosphorus moves slowly from deposits on land and in sediments, to living organisms, and than much more slowly back into the soil and water sediment.

The phosphorus cycle is the slowest one of the matter cycles that are described here.

Phosphorus is most commonly found in rock formations and ocean sediments as phosphate salts. Phosphate salts that are released from rocks through weathering usually dissolve in soil water and will be absorbed by plants. Because the quantities of phosphorus in soil are generally small, it is often the limiting factor for plant growth. That is why humans often apply phosphate fertilizers on farmland. Phosphates are also limiting factors for plant-growth in marine ecosystems, because they are not very water-soluble. Animals absorb phosphates by eating plants or plant-eating animals.

Phosphorus cycles through plants and animals much faster than it does through rocks and sediments. When animals and plants die, phosphates will return to the soils or oceans again during decay. After that, phosphorus will end up in sediments or rock formations again, remaining there for millions of years. Eventually, phosphorus is released again through weathering and the cycle starts over.

A schematic representation of the phosphorus cycle:

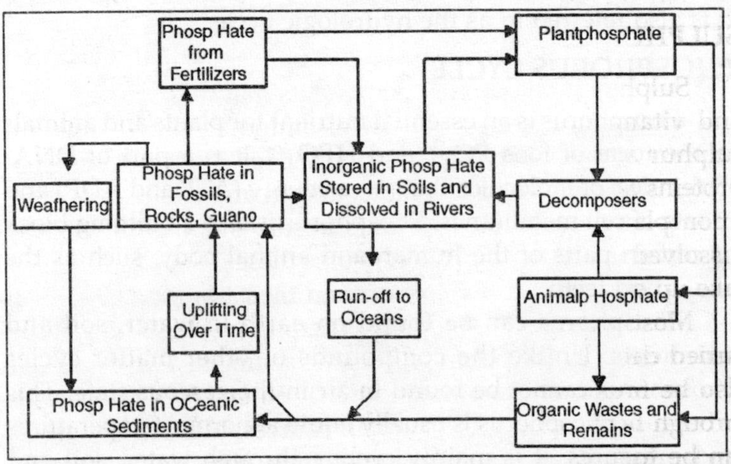

Humans also cause environmental problems by interfering with the phosphorus cycle. We remove phosphates on certain locations, to apply it on farmland as fertilizers. On these locations the phosphate supply is moving, causing phosphate levels of surrounding land-soils and groundwater to become too high.

On farmland where phosphates are applied as fertilizers, plants do not absorb all phosphate. The phosphates end up in water and stream towards lakes and reservoirs, where they cause a phenomenon called eutrophication. Eutrophication means that the water is so rich in nutrients that it causes certain water plants, such as green algae, to grow extensively. As a result the oxygen supply in the water depletes, causing fish and other oxygen-dependent organisms to die and bacteria that are not oxygen dependent (anaerobic) to take over.

When people practise extensive wood chopping in forest areas, such as the tropical rain forest, phosphate supplies deplete, because most of the phosphate can be found in the ground under the trees in such areas. When the trees are removed the rain washes phosphates away, causing the ground to become unproductive.

Phosphate supplies are restored very slowly, because most phosphate in soils and water comes from weathering rocks and oceanic sediments.

SULPHUR CYCLE

Sulphur is one of the components that make up proteins and vitamins. Proteins consist of amino acids that contain sulphur atoms. Sulphur is important for the functioning of proteins and enzymes in plants, and in animals that depend upon plants for sulphur. Plants absorb sulphur when it is dissolved in water. Animals consume these plants, so that they take up enough sulphur to maintain their health.

Most of the earth's sulphur is tied up in rocks and salts or buried deep in the ocean in oceanic sediments. Sulphur can also be found in the atmosphere. It enters the atmosphere through both natural and human sources. Natural recourses can be for instance volcanic eruptions, bacterial processes, evaporation from water, or decaying organisms. When sulphur enters the atmosphere through human activity, this is mainly a consequence of industrial processes where sulphur dioxide (SO_2) and hydrogen sulphide (H_2S) gases are emitted on a wide scale. When sulphur dioxide enters the atmosphere it will react with oxygen to produce sulphur trioxide gas (SO_3), or with other chemicals in the atmosphere, to produce sulphur salts. Sulphur dioxide may also react with water to produce sulphuric acid (H_2SO_4). Sulphuric acid may also be produced from demethylsulphide, which is emitted to the atmosphere by plankton species. All these particles will settle back onto earth, or react with rain and fall back onto earth as acid deposition. The particles will than be absorbed by plants again and are released back into the atmosphere, so that the sulphur cycle will start over again.

A schematic representation of the sulphur cycle:

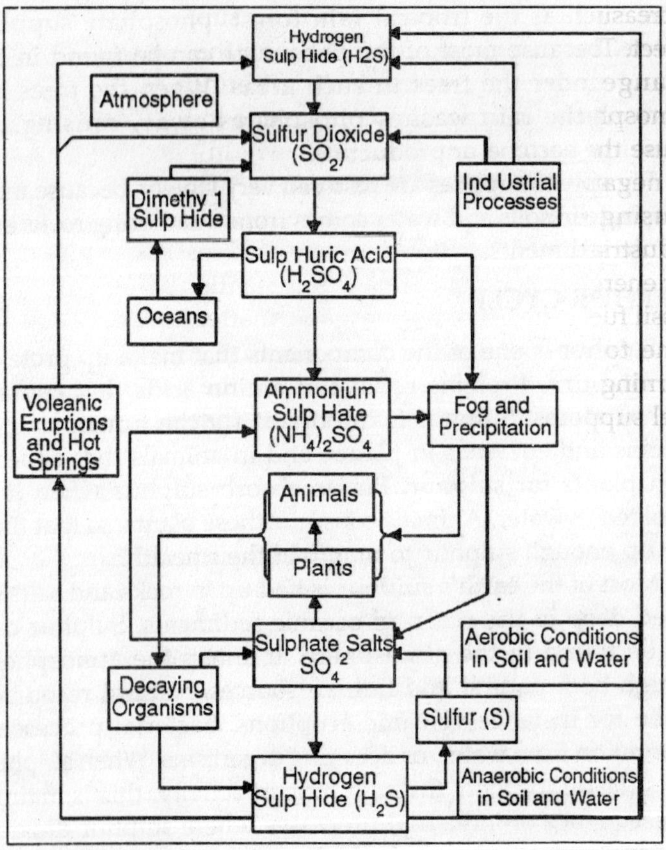

Humans cause environmental problems by influencing the

carbon cycle in two ways. Firstly, the removal of forests has caused depletion in plants and trees that absorb carbon dioxide. Humans cause great carbon emissions to the atmosphere, which take place during industrial processes, such as coal and oil combustion. These processes serve the generation of energy. As a result of these effects the carbon dioxide levels in the atmosphere have greatly increased. Between 1870 and 1990 the carbon dioxide emissions to air have increased 25%. The carbon dioxide levels in air are now so high, that the uptake by plants and oceans is not fast enough.

Carbon dioxide is a greenhouse gas. This means that increasing carbon dioxide levels in air support the greenhouse effect. The greenhouse effect.is a term that is used for climate change on earth, caused by greenhouse gases in the atmosphere. Too much carbon dioxide in the atmosphere can cause the earth to warm up. This will in time cause all kinds of negative effects, such as melting of ice from the Arctic causing floods on other continents. Secondly, during the industrial revolution humans have used fossil fuels extensively for energy generation, as has been mentioned before. Because fossil fuels consist of dead organic matter it takes a very long time to restore the supplies. Because humans have been burning fossil fuels extensively over the past decades fossil fuel supplies are in danger of being exhausted.

Chapter 2

Concept of Microbiology

OVERVIEW

Microbiology is the study of microorganisms, that is the organisms which are of microscopic dimensions. These organisms are too small to be clearly perceived by the unaided human eye. If an object has a diameter of less than 0.1 mm, the eye can not perceive (or more correctly resolve) it at all, and very little detail can be perceived in an object with a diameter of I mm. Roughly speaking organisms with a diameter of 1 mm or less are microorganisms and fall into the broad domain of microbiology. Since most microorganisms are only a few thousandths of a millimeter in size. they can only be seen with the aid of microscope.

As a direct consequence of the invisibility of micros to the naked eye and the need for special techniques to study them, microbiology was the last of the three major divisions in biology (the other two are botany and zoology to develop. Microbiology is the study of organisms that are too small to be clearly seen by the unaided eye. Roughly speaking, organisms with a diameter of 1 mm or less which can not be seen by the human eye unaided, are called microorganisms. These include protozoa, algae, fungi and bacteria Viruses are ultramicroscopic and have an obligate parasitic relationship, but for practical purposes these still come under the domain of microbiology.

At present, there is general agreement to include five major groups as microorganisms: the subdivisions of virology, bacteriology, mycology, phycology and protozoology (the studies of viruses, bacteria, fungi, algae and protozoa respectively). Traditionally, all organisms had been included under the disciplines of either botany or zoology. Bacteria, algae and fungi have in the past been considered to be the part of the Plant Kingdom while protozoa have been included in the Animal Kingdom. This view cannot be supported from a taxonomic standpoint. Since microbiology encompasses the study of groups of organisms in all three of these divisions of biology it may be argued that it covers a greater biological diver itv than the other two divisions.

Although microorganisms have existed for a long time, their existence was unknown until the invention of the microscope in the 17th century. The year 1674, marks the birth of microbiology when Antony van Leeu wenhoek a Dutch cloth merchant, looked at a drop of lake water through a glass lens which he had ground. What he observed through this simple magnifying lens was an amazing sight since that was perhaps the first time that man ever had a glimpse of the world of the microbes. Subsequently, in a series of letters to the Royal Society of London, Leeuwenhoek describ ed a variety of microorganisms such as .protozoa, algae, yeast and bacteria and the description was so precise that it is now possible to assign them into specific genera without any additicnal descrip-tion. Leeuwenhoek had little formal education but his keen interest in nature made him to examine a variety of materials. Glass grinding and preparation of lense? was his hobby and this led him to the assembly of about 400 simple microscopes. The earliest microscope that he constructed consisted of a spherical lens mounted on two plates

This process of unilinear inheritance of the transferred gene continues leaving a trail of nonmotile celled colonies until the gene is lost for reasons not well known. Using single cell studies, this mechanism was confirmed as abortive transduction.

Later in 1956, Ozeki also observed the same type of phenomenon in' transduction of nutritional characters in Salmonella diagrammatically summarises linear inheritance as it occurs in abortive transduction.

SCOPE

The aim of this component on microbiology is to enable you to understand the concepts underlying the biology of microbes. We will achieve this aim by examining aspects of their biology that differentiate microbes from larger organisms. We revise and build on previous material on the structure and evolution of the cell.

This component finishes with some extension material on competition. By the end of the unit you will be expected to apply these concepts to any practical situation and develop a response to a problem or issue that involves microbes.

This means you need to be able to recall, integrate and apply information and concepts from this part of the unit. Please be aware that the structure of this electronic learning resource does not strictly follow the lectures, though the underlying concepts are exactly the same.

CONCEPTS

The concepts covered are:

- The enormous functional diversity within microbes, despite morphological similarity.
- The different methods microbes use to acquire energy and oxygen.
- The potential for microbes to increase the size of their populations.
- The mechanisms underlying microbial competition and the consequences of competition.

Other concepts in microbiology useful for understanding aspects of environmental microbiology are:

- Growth Rates of bacteria.
- Elemental composition and structure of bacterial cells
- General physiological functions of these elements in cells.
- General characteristics of bacteria (haploid cells, high growth rates, binary fission, biochemical diversity, high variation rate, grow in extreme environments, can survive via sporulation, etc).
- Generalized growth physiology (response to

environmental factors, growth in extreme environments, oligotrophy, etc.) and kinetics of growth (lag phase, log phase, stationary phase and decline or death phase of growth).

- Main groups of microorganisms in terms of physiology - aerobic, anaerobic, aerotolerant, facultative anaerobe, obligate anaerobe,
- *Groupings by trophic type*: Photoautotrophs, photoheterotrophs, chemoautotrophs, chemoheterotrophs, and mixotrophs
- General patterns of metabolism - energy reactions to "building block reactions", to polymerization to cellular structures.

CLASSIFICATIONS

In general, a classification is a method for organizing information. Human beings classify things spontaneously. For example, after seeing enough examples of chairs, we form a classification in which any given object is either a chair or a non-chair. Given a new object we do remarkably well at deciding whether it is a chair (in essence, whether or not it is practical and socially acceptable sit on it). Although there is little scientific value in this particular classification, it illustrates their ubiquity.

CLASSIFICATIONS GROUP

In general, classifications group similar things together. This definition is necessarily vague; there are many reasonable ways of defining similarity, and hence many alternative classifications for the same things.

In the example of chairs, we considered the relationship (similarity) of objects to an abstract definition of chair. This definition was based on a potential function of the object. In choosing to classify by "chairishness", we assign less importance to other properties—Colour, hardness, age, etc.—that can vary without altering the utility of the object as a chair.

There is no "right way" to classify things. For example, words can be classified:

- By their spelling (as in a dictionary).
- By the spelling of a foreign language synonym (as in a bilingual dictionary).
- By their meanings (as in a thesaurus).

The intended use of a classification is an integral part of deciding what properties it should be based upon. Here are some other classifications that you have encountered: the Library of Congress book classifications (organized by subject); the fiction section in a library (usually organized by the author's last name); and a food store (organized primarily by storage temperature).

An early biological classification is implicit in the ancient Greeks' definition of the three kingdoms: animal, vegetable and mineral. This classification split animate matter into (what we would now call) plants and animals. More detailed classifications also existed. For example, Aristotle grouped humans and birds together, since they walk on two feet.

Classifications are Often Hierarchical

A key property of classifications is that they can be nested within one another, creating an hierarchy. Thus, any group within a classification can be split in still greater detail. For example, objects that have been classified as chairs might be subdivided into stools, rocking chairs, recliners, etc.

Similarly, listings in the "yellow pages" of a telephone directory are first classified by the product or service, and then within each of these categories they are further classified alphabetically.

There is no limit to the depth of a hierarchical classification. Most only have a few levels, but there are some that are quite deep.

When alphabetical order is considered carefully, it is seen that the first level of classification deals only with the first letter of each word. Then all words starting with the same first letter (for example "c") are categorized by their second letter, adding a second level to the alphabetical classification. This process continues one letter at a time until all words have been distinguished.

Biological Classification

Linnaeus defined the biological classification system that we still use for plants and animals, and, with relatively minor modifications, for fungi and microorganisms. It is a hierarchical system that starts with a few categories at the highest level, and further subdivides them at each lower level.

The levels in the hierarchy were given names that are already familiar to you:

Domain Kingdom Phylum Class Order Family Genus Species.

Using this system, a fruit fly might be classified as:

Eucarya Animalia ArthropodaInsecta Diptera Drosophilidae *Drosophilamelanogaster.*

This means that this fly is the species melanogaster in the genus *Drosophila* in the Family Drosophilidae in the Order Diptera in the Class Insecta in the Phylum Arthropoda in the Kingdom Animalia in the Domain Eucarya. In the Linnaean system, to uniquely name a species it is necessary to supply both the genus and species,*Drosophila melanogaster.*

In the case of animals, Linnaean classifications often reflect our intuitive sense of similarity. Thus in the following classifications species 2 and 3 have more levels of classification in common than either has with species 1. This is a reflection of the greater similarity of the fly (*Drosophila melanogaster*) and the mosquito (*Aedes aegypti*), than of fly and lobster (*Homarus americanus*) or mosquito and lobster.

Table. Classifications of Three Species

	Species		
Level	1	2	3
Domain	Eucarya	Eucarya	Eucarya
Kingdom	Animalia	Animalia	Animalia
Phylum	Arthropoda	Arthropoda	Arthropoda
Class	Crustacea	Insecta	Insecta
Order	Decapoda	Diptera	Diptera
Family	Caridea	Drosophilidae	Nematocera
Genus	Homarus	Drosophila	Aedes
Species	americanus	melanogaster	aegypti

The shared levels of classification can be accentuated by slightly reformatting the data in the table.

Table. Classifications of Three Species

Level	**Species** **1**	**2**	**3**
Domain		Eucarya	
Kingdom		Animalia	
Phylum		Arthropoda	
Class	Crustacea		Insecta
Order	Decapoda		Diptera
Family	Caridea	Drosophilidae	Nematocera
Genus	Homarus	Drosophila	Aedes
Species	americanus	melanogaster	aegypti

NATURAL SYSTEM

Although there are no right and wrong ways to classify things, when the natures of objects are defined by a common history then there is a "natural" way to classify them. For most objects, their natures are largely independent of their histories; but organisms are products of their genetic history.

Similarity in Organisms

For millennia it has been appreciated that organisms give rise to new individuals of the same kind (i.e., same species). Thus, humans give rise to more humans, dandelions give rise to more dandelions, *Escherichia coli* gives rise to more E. coli, and so forth. In essence, like begets like begets like begets like *etc. ad nauseam*. The continuity of a species is a consequence of the genes passed from each generation to the next. This is a static view of the biological world; without other processes, life could not change.

Species are Dynamic

In the 19th century, biologists began to recognize that species are dynamic, not static; that is, species change over time. Several discoveries necessitated this view and overcame seemingly conflicting observations. For example, the "like begets like" view does not accommodate the variation evident among individuals of a species. Also, the discovery of fossils indicated that ancient life forms were not the same as those alive today.

Some of the fossils could even be organized into progressions, with later forms being similar, but not identical, to earlier forms. The simplest modification to the static view would be to suggest that like begets similar. At the same time, work in geology was revealing that the Earth was old—very much older than had been previously believed. This meant that even if changes between successive generations were small, there had been time for changes to accumulate into the large divergences that distinguish species.

The impact of this new perspective is profound. In the earlier static view, if we followed members of a species (call it A_0) over time, their descendants and their descendants' descendants, and so on, would belong to the same species, forever, regardless of the duration of the intervening interval:

$$A_0 \rightarrow A_0 \rightarrow A_0 \rightarrow A_0 \rightarrow \ldots \rightarrow A_0 \rightarrow A_0$$

In contrast, the dynamic view of species implies that, given sufficient time, naturally occurring changes would accumulate, ultimately changing the lineage to such an extent that a descendent would be a member of a new species, A_1, that is distinct from that of its distant ancestors:

$$A_0 \rightarrow A_0' \rightarrow A_0'' \rightarrow A_0''' \rightarrow \ldots \rightarrow A_1' \rightarrow A_1$$

This realization was the motivation for Darwin's title: *On the Origin of Specie*[7].

This could be represented diagramatically as:

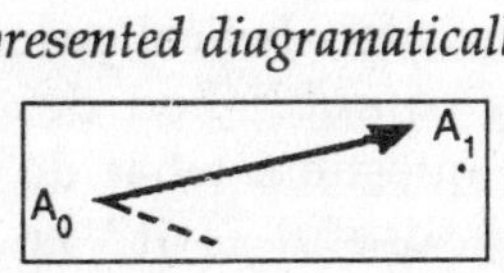

Consequence of Ancestry

Consider the consequences of species A_0 giving rise to two lineages that are prevented from further interbreeding. With sufficient time, the first lineage would change enough to be called a new species, A_1 (as above), while in the same interval of time, the second group would also have changed enough to be a new species. Because the two lineages have accumulated changes independently (there was no interbreeding), most of the changes in the groups will be different. Thus the second group will have become a species that is not the same as either A_0 or A_1, so it needs a name of its own, A_2:

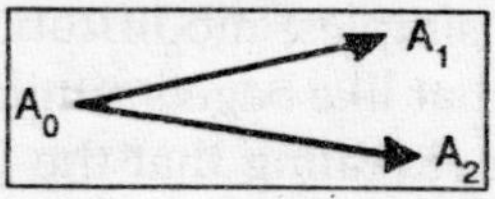

Species A_1 and A_2 must each bear residual similarity to A_0 (their most recent common ancestor), so they must also be similar to each other. Most similarities between present-day species are due to their common ancestry. To first approximation, the more recent their common ancestry, the more similar two species will be (the more properties they will share).

Turning this around, the most similar species today tend to be those that share the most recent common ancestry. More distinct species tend to have more remote common ancestry. Darwin (and others) saw that once we accept this perspective, then we inevitably continue until we have accepted the common ancestry of all life.

A Natural System of Classification for Organisms

As noted above, there is no absolute rule for designing a classification; we can classify things by any criterion we choose. This is true in classifying organisms, as well. We can classify by size, shape, Colour, alphabetical order (in any language), or even chronology of published descriptions. There is however a practical question: what do we learn from our classification? How useful is it? Is there a preferred classification of organisms from which we can learn more than we can from alternative classifications?

If we classify by an arbitrary criterion, then knowing an organism's placement within the classification can be remarkably uninformative. For example, if we classify bacteria on the basis of shape (i.e., rod, coccus, spiral, etc.), then knowing an organism's classification tells us little more than its shape. We cannot begin to guess whether it is pathogenic, whether it makes food taste good, or whether it enhances the growth of our lawn.

In the case of organisms, we can do much better than an arbitrary classification. Ideally, we wish to base our classification on the relationships of the organisms, that is, on their genealogy or phylogeny. This is uniquely useful because knowing that two organisms are closely related tells us that they will have many properties in common. Knowing historical relationships has more predictive potential (more applicability to properties that have not yet been measured) than does an arbitrary classification.

Darwin referred to a classification based on evolutionary history as a natural system. In recognizing the parallel between phylogenetic closeness and species similarity, Darwin explained why the Linnaean system had been so successful for macroscopic plants and animals: they had been placed into the hierarchical groups of the Linnaean classification on the basis of similarities, and because these similarities reflected the evolutionary closeness of the respective species, the classification tended to reflect their historical relationships. In general, when similarities due to common ancestry are easy to recognize and measure, a classification of the organisms tends to approximate a natural system.

The original successes of the Linnaean system were largely limited to complex plants and animals. In these cases there are numerous features that can be seen and compared between organisms. In addition, many of the features are sufficiently complex that it is safe to assume that they were only invented once, so that similarities really are the result of inheritance from a common ancestor that also possessed the shared feature.

CLASSIFYING MICROORGANISMS

Determinative Microbiology

The prokaryotes and eukaryotic protists were more difficult to classify. They are much simpler, and there are few obvious features that are sufficiently widespread to permit meaningful measurements of similarity between distantly related organisms. Some of the features that have been used to classify prokaryotes include: Gram stain (cell wall structure) Mole per cent G+C in the genome Growth temperature Ability to form heat stable spores Electron acceptors for respiration (if any) Photosynthetic ability

Motility Cell shape Ability to use various carbon and nitrogen sources Special nutritional requirements (e.g., vitamins) Leading microbiologists in the 20th century (e.g., Stanier and van Niel) realized and acknowledged that classifications based upon features such as these were not natural systems, but they also realized that no satisfactory alternatives were available. Consequently, the resulting classification of bacteria (excluding cyanobacteria, which were classified with the eukaryotic algae) was codified in a volume called Bergey's Manual of Determinative Bacteriology. In doing so, it was recognized that the classification could be used to accurately identify organisms that had already been characterized. However, given a new isolate with the following properties: Gram-variable 32% G+C genome thermophilic (85 C optimum) nonspore-forming anaerobic nonphotosynthetic nonmotile rod-shaped prokaryote grows on H_2, CO_2 and N_2 no special growth factors required microbiologists were unable to predict its phylogeny or additional properties. That is, it was not possible to say much about a new organism other than what was directly measured.

MOLECULAR PHYLOGENY

Although the morphologies and physiologies of prokaryotes are much simpler than those of eukaryotes, there is a large amount of information in the molecular sequences of their DNA, RNAs and proteins. [15] Thus, it is possible to use molecular similarities to infer the relationships of genes, and, by extension, to learn the relationships of the organisms themselves.

DNA-DNA Hybridization

One important technique for comparing prokaryotes at

the molecular level is DNA-DNA hybridization. In this test, the genomic DNA from one species is mixed with the DNA from a second species and the similarity of the DNAs is reflected in the extent to which strands of DNA from one organism anneal with strands of DNA from the other organism.

The sensitivity of DNA-DNA hybridization declines rapidly as the organisms become more diverged, limiting the method to characterization of closely related strains, species and genera. In addition, testing the relationships of a new organism can require many hybridizations. If no close relative is found by this test, we have only learned what the new organism is not—whereas we wish to know what it is.

Phylogenies Based on Specific Genes

DNA-DNA hybridization gives a measure of relatedness across the whole genome. As mentioned above, the average similarity falls off rapidly when looking at more diverged species. In principle, the range accessible to molecular analysis could be increased by looking at specific genes with above average conservation.

There are several important requirements if we wish to use a gene phylogeny to infer organismal relationships:

- The gene must be present in all organisms of interest. Thus, to infer relationships that span the diversity of prokaryotes (or life), we must look at the central (universal) cellular functions. Examples include genes whose products function in replication, transcription, or translation—the processes constituting the "Central Dogma" of molecular biology.
- The gene cannot be subject to transfer between species (lateral transfer). Since we wish to infer organismal relationships, if a gene is transferred, then the gene history is not the same as the organismal history. If a gene performs a central function, an organism is unlikely to acquire a copy by lateral transfer, since the organism must already have a functional copy to be alive.

- The gene must display an appropriate level of sequence conservation for the divergences of interest. If there is too much change, then the sequences become randomized, and there is a limit to the depth of the divergences that can be accurately inferred. If there is too little change (if the gene is too conserved), then there may be little or no change between the evolutionary branchings of interest, and it will not be possible to infer close (genus or species level) relationships.
- The gene must be sufficiently large to contain a record of the historical information. Thus, although transfer RNA (tRNA) genes are present in all species, they are too small (about 75 nucleotides) to provide an accurate sample of evolutionary history.

Ribosomal RNA Genes and Their Sequences

To infer relationships that span the diversity of known life, it is necessary to look at genes conserved through the billions of years of evolutionary divergence. An example of genes in this category are those that define the ribosomal RNAs (rRNAs). Most prokaryotes have three rRNAs, called the 5S, 16S and 23S rRNA.

Ribosomal RNAs in Prokaryotes

Name[a]	Size (nucleotides)	Location
5S	120	Large subunit of ribosome
16S	1500	Small subunit of ribosome
23S	2900	Large subunit of ribosome

[a] The name is based on the rate that the molecule sediments (sinks)in water. Bigger molecules sediment faster than small ones.

The 5S has been extensively studied, but it is usually too small for reliable phylogenetic inference. The 16S and 23S rRNAs are sufficiently large to be quite useful.

The extraordinary conservation of rRNA genes can be seen in these fragments of the small subunit rRNA gene sequences from organisms spanning the known diversity of life:

As a graduate student at the University of Illinois, Bernadette Pace used the annealing of rRNA with genomic DNA to measure the similarity of rRNAs in various species. These experiments demonstrated that rRNA-based methods are applicable to directly comparing a broader range of organisms (i.e., spanning greater phylogenetic distances) than is whole genome DNA-DNA hybridization.

However, as with DNA-DNA measurements, it was necessary to have DNA and/or RNA from each species of interest.

If relationships were Analysed by comparing sequence data, rather than hybridizing the molecules, one could infer relationships without having all of the molecules in hand (only the sequence data from previous studies are necessary). This was already being done with protein sequences.

Carl Woese recognized the full potential of rRNA sequences as a measure of phylogenetic relatedness. He initially used an RNA sequencing method that determined about 1/4 of the nucleotides in the 16S rRNA (the best technology available at the time). This amount of data greatly exceeded anything else then available. Using newer methods, it is now routine to determine the sequence of the entire 16S rRNA molecule. Today, the accumulated 16S rRNA sequences (about 10,000) constitute the largest body of data available for inferring relationships among organisms.

Molecular Phylogenies can Reflect Genealogy and Amount of Change By comparing the inferred rRNA sequences (or those of any other appropriate molecule) it is possible to estimate the historical branching order of the species, and also the total amount of sequence change. An example of a 16S rRNA-based phylogenetic tree showing the three (identified) Domains of life—Bacteria, Archaea and Eucarya—is below.

In this tree, lineages diverge from a common ancestral lineage on the far left. The lengths of the individual lines reflect the amount of sequence change (note that some lineages have modified the gene sequence substantially more than others, and thus have accumulated longer total branch lengths).

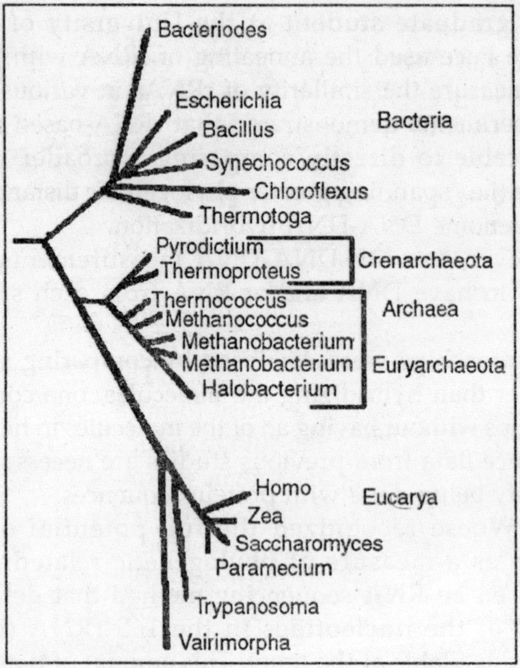

EUCARYA (EUKARYOTES)

Major Divisions of Eucarya

The Eucarya are usually classified into four kingdoms: plants, animals, fungi and protists. The first three of these correspond to phylogenetically coherent groups as well. However, the eukaryotic protists do not form a group, but rather are comprised of many phylogenetically disparate groups (including slime molds, multiple groups of algae, and many distinct groups of protozoa). Just as molecular analyses where required to see the natural relationships among prokaryotes, they are also allowing us to infer relationships among these non-plant, non-animal, non-fungus eukaryotes.

Many of the groups of fungi are important. In addition to providing yeast for making bread and fermenting sugar to alcohol, the fungi include the causative agents of such conditions as blight, ergot, rot, rust, smut, wilt, ringworm and athlete's foot. Although it might seem that it is straightforward to identify an organism as a fungus, this is not always the case. The unambiguous identification of the opportunistic pathogen Pneumocystis (which has not been grown in culture) as a fungus required analysis of its ribosomal RNA.

ARCHAEA

Major Groups within the Archaea

The major physiological groups within the Archaea (informally, archaes) are extreme thermophiles (including thermoacidophiles), sulfate reducers, methanogens, and extreme halophiles. These physiological groups do not correspond to the phylogenetic groups. This is largely because the methanogens have given rise to multiple other groups, including the halophiles, the sulfate reducers, and at least two other lineages of nonmethanogenic thermophiles.

The thermophilic members of the Archaea include the most thermophilic organisms cultivated in the Labouratory, some having optimal growth at >100 *f*C. The aerobic thermophiles are also acidophilic; they oxidize sulfur in their environment to sulfuric acid, creating acid hot springs. The hot acidic waters can dissolve limestone and some other minerals out of the surrounding rock matrix, creating boiling mud pots of the insoluble residue from the rock.

The extreme halophiles include the most salt tolerant organisms known. They are aerobic or microaerophilic. They have also invented a form of photosynthesis that does not use chlorophyll.

The sulfate-reducing Archaea are now known to be a wide spread group. One of the reasons that this group has received attention is that sulfate-reducing organisms are responsible for "souring" oil wells (increasing their sulfur content). In addition to increasing the sulfur emissions when the oil is burned (or increasing the cost of refining it), sulfide attacks the metal in well casings and pipelines.

Methanogens are strict anaerobes, yet they gave rise to at least two separate aerobic groups: the halophiles and a thermoacidophilic lineage. Methanogens produce most of the methane in the Earth's atmosphere. Methane is a significant green-house gas (i.e., it is a cause of global warming). Much of this methane production is associated with activities for which humans are at least partially responsible. This includes the flooding of rice paddies and the raising of domestic cattle.

Archaea are Prokaryotes

The Archaea are not eukaryotes, which makes them prokaryotes. However, the fact that they are related to Eucarya, not to the other prokaryotic organisms (the Bacteria), means that prokaryotes are not a natural group.

This is difficult to fully appreciate, since we tend to accord a special status to eukaryotic cells. It is perhaps most easily understood by considering the ability to learn about one group from another:

If we wonder about a biosynthetic reaction in a eukaryote, it makes more sense to use a member of the Archaea as a model than to use a member of the Bacteria. Even though much more money is being spent on analysis of eukaryotic genomes than prokaryotic genomes, due to their smaller size, some prokaryotic genomes will be completed first. Of these, it appears that we will learn more about human genes from the archaeal genomes than from the bacterial genomes.

If we wonder about transcription initiation in *Thermococcus celer* (an archaeon), should we look at what is known about transcription in *E. coli*(another prokaryote) or in yeast (a eukaryote)? It is now known that the transcription apparatus of *T. celer* seems to be much more similar to that of yeast than to that of *E. coli.*

BACTERIA

Some Major Groups within the Bacteria

Some major phylogenetic divisions within the eubacteria and representative genera are listed below. Additional divisions continue to be found. The groups (and some of the genera) can be seen in the last page of the handout. To illustrate the scattered distribution of some physiological characteristics, they are marked on the corresponding genera: photosynthetic with an asterisk (*); gliding motility with a bar (|), lack of a cell wall with a paragraph mark (¶), and thermophilicity by a section mark (§).

Hydrogen oxidizers(oxygen reducers)	Aquifex
Thermotogales	*Thermotoga, Fervidobacterium*
Thermodesulfobacterium group	*Thermodesul fobacterium*
Green nonsulfur	*Chloroflexus**¶, *Herpetosiphon, Thermomicrobium*
Deinococcus group	*Deinococcus, Thermus*
Thermodesulfovibrio group	*Thermodesulfovibrio*§
Synergistes group	*Synergistes*
Low G+C Gram positive	*Bacillus, Clostridium, Eubacterium, Heliobacterium**, *Lactobacillus, Mycoplasma, Spiroplasma*
High G+C Gram positive	*Bifidobacterium, Mycobacterium, Propionibacterium, Streptomyces*
Cyanobacteria	*Oscillatoria**¶, *Prochlorococcus**, *Synechococcus**, chloroplasts*
Planctomycetales	*Planctomyces*
Chlamydiales	*Chlamydia*
Green sulfur	*Chlorobium**
Cytophaga group	*Bacteriodes, Cytophaga*¶, *Flexibacter*¶, *Flavobacterium, Rhodothermus*
Fibrobacter group	*Fibrobacter*
Spirochetes	*Borrelia, Leptonema, Spirochaeta* (including *Spirochaeta* sp. str. Antarctic), *Treponema*
Proteobacteria (Purple bacteria):	
alpha subdivision	*Agrobacterium, Anaplasma, Rhodobacter, Rhodospirillum**, *Rickettsia*, mitochondria
beta subdivision	*Neisseria, Rhodocyclus**
gamma subdivision	*Beggiatoa*¶, *Chromatium**, *Escherichia, Haemophilus, Legionella, Pseudomonas, Salmonella, Vibrio, Yersinia*
delta/epsilon subdivision	*Bdellovibrio, Campylobacter, Desulfovibrio, Helicobacter, Myxococcus*¶, *Wolinella*

Chlorophyll Based Photosynthesis

Five of the major phylogenetic groups of Bacteria listed

above include photosynthetic genera. Given the complexity of the process, it is unlikely to have been invented multiple times. Instead, it seems much more likely that it originated early in eubacterial evolution, and has been lost (or gained by lateral transfer) in many lineages.

Some of the branches that diverge earliest in the Bacterial domain (notably the hydrogen oxidizers and the Thermotogales) have no known photosynthetic members. This suggests, but does not prove, that chlorophyll-based photosynthesis was invented after these groups diverged from the rest of the Bacterial lineages.

Oxygenic Photosynthesis

Although chlorophyll-based photosynthesis is widely distributed among the Bacteria, those that perform oxygenic photosynthesis form a phylogenetically monolithic group: all members of the cyanobacterial group share this physiology, and no organisms outside of the cyanobacteria have it.

Photosynthetic eukaryotes all contain chloroplasts (more generally called plastids). One of the great triumphs of molecular phylogenetic analyses has been the demonstration that the chloroplast genome is of bacterial origin. That is, the chloroplast was once a bacterium (more specifically, a cyanobacterium), and, after incorporation into the eukaryotic host cell, it has degenerated into a much simpler cellular organelle (it is an "endosymbiont'). Because of its loss of autonomy, the organelle is maintained by the host cell. In return, it allows the host cell to harvest light energy and to convert carbon dioxide and water into useful organic compounds.

The cyanobacterial lineage is responsible for essentially all biologically produced oxygen, whether by eukaryotic plants and algae, or by free-living cyanobacteria.

ORGANIZING DATA ON A TREE

Thermophilia

Thermophilia is represented in, or is the exclusive phenotype of, all of the deepest groups in the Archaea and the Bacteria. The simplest explanation is that the ancestral organisms were also thermophilic. This is appealing since it is believed that the early Earth was hot compared to its condition today.

Anaerobic Origins

Many of the deeply branching lineages are also anaerobic or microaerophilic (i.e., they use little or no oxygen). This is consistent with an early environment with less oxygen (certainly < 1%) than the present atmosphere. The geological evidence supports this low concentration. Higher oxygen levels had to await the invention of oxygenic photosynthesis.

Genetically-Defined Properties that have Changed Relatively Often Some properties of organisms are volatile; they seem to be frequently invented and/or lost.

Cell shape	The basic bacterial shapes (especially rods and cocci) are often intermixed within groups of related species. The most interesting exception is the spirochetes, most which fall in a single group.
Aerobiosis	The aerobic life-style is distributed (discontinuously) throughout the prokaryotic groups. This is in some respects puzzling, since there were not large amounts of atmospheric oxygen until about 2 billion years ago. It is highly unlikely that the separation of all of the different lineages that include aerobes is more recent than this date. Therefore one of the exciting questions confronting microbiologists is how did some many different lineages invent or acquire the required genes.
Ability to use various	Within several of the bacterial *donors electron and acceptors* groups, closely related organisms can use quite different compounds as electron donors or acceptors.

	For example, sulfur oxidation is scattered throughout most of the major prokaryotic lineages.
Chlorophyll-based photosynthesis	Although it is almost certain (given its complexity) that chlorophyll-based photosynthesis was only invented once, the property is found broadly scattered throughout the Bacteria. It appears that it was invented very early is eubacterial evolution, and has been lost many times.
Motility	One of the most surprisingly variable properties of various lineages is motility. Although it is easy to understand how motile organisms could have non-motile relatives (it could just be the loss of a single gene), it has also been found that gliding motility is found at several different places in the tree as well.
G+C content	Within very closely related species, the mole per cent G+C in the genome will tend to be within a few per cent. However, as one considers progressively deeper groups in the tree, the variation can become quite large. The net effect of this is that having similar G+C contents cannot be taken as an indication of a close relationship, but having a very different base content means that the organisms are not closely related.

Genetically-Defined Properties that have Changed Relatively Rarely

Not all cellular properties display as much variability. These properties are among those for which knowledge of an organism's phylogenetic position is most predictive.

Cell wall structure	A muramic-acid-containing cell wall indicates that the organism is a member of the Bacteria, since this structure is not found among Archaea or Eucarya. Some of the more specific cell walls are also quite diagnostic. One example would be the Gram positive cell wall. If an ultrastructural study shows a new organism to have a typical Gram positive cell wall, then the organism will almost certainly belong in this phylogenetic group. In addition, a cell known to be within one of the major groups of mycoplasmas will almost certainly lack a cell wall.
Membrane lipids	Structure of lipids is particularly valuable in distinguishing bacterial membranes (mostly fatty acids) from the archaeal membranes (mostly phytyl alcohols). There are other lipids that have proven valuable for recognizing other specific groups, particularly among the eubacteria. One of the most interesting applications of this information is to characterize the organisms that contributed to ancient oil deposits. This area is undergoing a recent renewal, with the introduction of better techniques and instrumentation.
Oxygenic photosynthesis	Although chlorophyll-based photosynthesis is widely distributed among Bacteria, only the cyanobacteria and chloroplasts (which are phylogenetically part of the cyanobacterial group) are able to extract the required electrons from

	water (thereby releasing oxygen). To date, all organisms known to perform oxygenic photosynthesis are members of the cyanobacterial group, and all free-living members of the phylogenetic group are able to perform oxygenic photosynthesis. It seems that members of the group are irreversibly committed to this mode of making a living.
Methanogenesis	All known methanogens belong to one of four identified related lineages within the Archaea. Although the methanogens have given rise to other organisms that are not methanogenic, it appears that methanogenesis has only been invented once.

MOLECULAR MICROBIAL ECOLOGY

One of the most dramatic contributions of molecular phylogeny to our understanding of microorganisms is its application to the area of microbial ecology.

Difficulty of Growth for Characterization

Traditional methods of microbial ecology require that organisms from an environment be cultivated in the Labouratory so that they can be characterized and identified. However, if the number of cells visible in an environmental sample are directly counted and compared with the number of cells successfully cultivated from the sample, it is found that anywhere from 90% to 99.9% of the cells defy cultivation, and therefore evade identification. Thus, most characterizations by cultivation methods have contained a large bias in an unknown direction.

In addition, if a new sequence is related to those from organisms with particular physiological properties, then those properties can be extrapolated onto the organism in the environment that was the source of the particular new sequence. Thus, it is even possible to learn something about the biochemical processes that might be going on in an environment, even without the ability to grow the organisms and make the measurements in the Labouratory.

The same methodologies are being applied to human and animal pathogens. These techniques have permitted the phylogenetic identification organisms that have defied cultivation and have provided the basis for rapid detection of specific organisms or groups of organisms in clinical samples.

Branching Order in Phylogenetic Tree

The branching pattern (often called branching order) shows the genealogy of the organisms. That is, it shows which species share more common ancestry than which others.

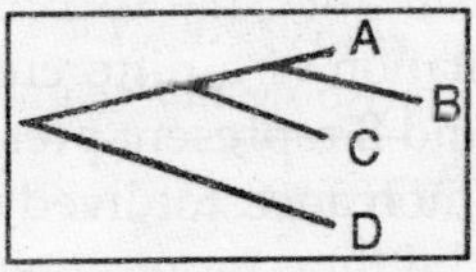

In the tree above, A and B share the most recent common ancestry. Thus, of the species in the tree, A and B are the most closely related.

The next most recent common ancestry is C with the group composed of A and B. Notice that the relationship of C is with the group containing A and B. In particular, C is not more closely related to B than to A. This can be emphasized by the following two trees, which are equivalent to each other:

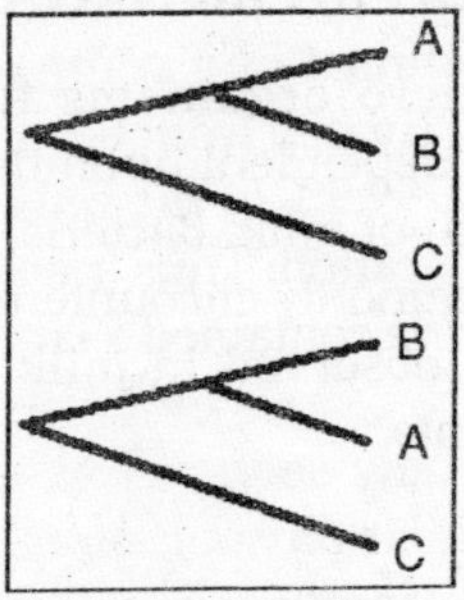

It is only a matter of taste which way it is drawn. In essence, in both versions A and B share a more recent common ancestor than do either of them with C. Thus it does not make sense for either A or B to be closer to C than is the other, since the alternative drawing would seem to imply the opposite.

Going back to the first drawing:

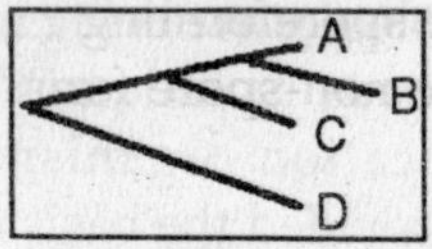

The most ancient common ancestry in this tree is that of D with the group composed of A, B and C.

Branch Lengths in a Phylogenetic Tree

In the above drawing, the length of the line leading from the A+B common ancestry to A is shorter than the line leading from the A+B common ancestry to B. This is intended to represent the accumulation of a different amount of change. Note that the both A and B represent present-day species, and their most recent common ancestor lived some specific period of time ago. That is, the time represented by the branches is the same, even though a different amount of change has accumulated.

To get unequal amounts of change in equal time periods, the rate of change must differ. There is no reason why rates of molecular change have to be either constant or equal in all lineages. When we examine molecular sequence data, we often find that some lineages have changed different average rates.

KINDS OF MICROORGANISMS

The types of micro organisms found in milk vary considerably, and are dependent upon the specific conditions associated with a batch of milk. Bacteria, yeasts, moulds, and bacteriophages are commonly encountered. Other viruses and Protozoa are seldom observed in milk products, except as occasional contaminants.

BACTERIA

Bacteria are the most common, and probably the most numerous of microorganisms with which the dairy processing industry is concerned.

They belong to four main groups:

- Cocci, usually gram positive.
- Gram positive non-sporeforming rods
- Gram positive sporeforming rods.
- Gram negative non-spore forming rods.

YEASTS

The yeasts most frequently encountered in milk and milk products act upon the lactose to produce acid and carbon dioxide. Yeasts are more commonly found in raw, cream during hot weather, but are potential contaminants throughout the year.

MOULDS

Moulds often grow in large concentrations and are visible as a fuzzy or fluffy growth. They are sometimes observed on the surface, of butter, old cream, khoa, or cheese. They are black, grey, green, blue or white. They discolour milk products and often produce undesir able, at times repulsive, odours and flavours. Moulds are essential in production of the certain kinds of cheese.

BACTERIOPHAGES

Bacteriophages are particularly obnoxious in starter cultures used for making cultured milk, butter and cheese. Phages will, kill the bacterial culture and the whole process of fermentation will be lost.

Microorganisms found in milk can also be described on the basis of the following characteristics;

- Biochemical activities.
- Temperature response.
- Ability to cause infection and disease.

BIOCHEMICAL ACTIVITIES

If allowed to stand under condition that permit bacterial growth, raw milk of a good sanitary quality will rapidly undergo a series of chemical changes. The principal change is lactose fermentation to lactic acid. This change is brought about by acid uric lactic organisms, especially Strepotococcus lactis and certain lactobacilli. These include two distinct biochemical types, homo-and heterofermentative. In homofermentation lactic acid is the major product of lactose fermentation.

Heterofermentative organisms, however, produce lactic, acetic, propionic, and some other acids, and some alcohols and gases such as CO_2 and H_2 Organisms continue to form lactic acid until the concentration of acid is itself too great for the organisms to remain live.

Microbacteria, micrococci, coliforn18, etc. also ferment lactose to lactic acid and other products. Many Clostriifiul1J species and, some yeasts such as Torula lactic, and Torula cremoris ferment lactose with acid and gas production. As the acidity continues to increase and reaches a pH of 4.7, it eventually causes a precipitation of casein. Organisms capable of metabolizing lactic and other acids develop especially acid uric, yeasts and moulds.

The acidity of milk is diminished and the alkaline products of protein decomposition such as amines, ammonia and the like are produced. This is accomplished by many species of the genera Bacillus, Clostridium, Pseudomonas, Proteus and numerous other forms.

The action of microorganisms does not involve fat as readily as it does lactose and protein. Lipolysis results from the action of lipase produced by bacteria such as Pseudomonas, Achromobacter and by some yeasts and moulds. Fat is hydrolysed to glycerol and fatty acids. Some of the fatty acids, for example, butyric and caproic acid give milk products, distinctive and usually rancid, odours and flavours.

Several microorganisms also bring about certain objection able changes in the milk which may not be deleterious to health. Rapines in milk is sometimes encountered. The milk become ropy or slimy and may be pulled out into long threads. It is produced by several organisms but the most important species is Alcaligenes viscolactis. A rapid fermentation of lactose in milk is sometimes observed and is known as stormy fermentation. This is brought about by Clostridium perfringens.

The curd become torn to shreds by the Vigourous fermentation and gas production. Several organisms have been isolated from milk which impart brilliant colours. Pseudomonas syncyanea imparts blue colour, pseudomonas synxantha yellow colour and Serratia marcescens red colour to the milk.

TEMPERATURE RESPONSE

Microorganisms found in milk can also be described according their optimum temperature for growth and heat resistance.

This is a very practical consideration since milk is preserved by employing low temperatures to prevent changes due to microbial activi1y'and by high temperatures to reduce microbial population and destroy pathogens. All the four types of microorganisms i.e. psychrophilic, mesophilic, thermophilic and thermoduric are found in milk.

Psychrophiles grow at temperatures just above freezing and at refrigeration tempratures. They produce a wide variety of spoilage defects.

The defects may result in the production bf many "off" flavours and odours. The most commonly encountered psychrophilic bacteria are members of the genera pseudomonas, Achromobacter, Vibrio, Flavobacterium and Alcaligenes, They arc killed in the pasteurization process, but are sometimes found in pasteurized milk. The contamin ation takes place after pasteurization from equipment, cans, bottles, and water.

The most important mesophilic bacteria are streptococci, lactobacilli and coliforms, which produce acid and gas and off flavours. They are killed in the pasteurization process

Thermophilic bacteria grow well at the temperature used in pasteurization, specially when the low temperature holding method is followed. Most thermophilic forms are found in two genera, Bacillus and Clostridium.

Thermoduric organisms are regarded as those which survive pasteurization but do not grow at pasteurization temperatures. The most common thermoduric bacteria are found in the genera Microbacterium, Corynebacterium, Micrococcus, Streptococcus and Bacillus.

Excessive numbers of thermoduric bacteria in milk make it difficult to meet the grading standard.

ABILITY TO CAUSE INFECTION AND DISEASE

Pathogenic organisms of both bovine and human origin have been isolated from milk. Milk, therefore, can serve as a carrier of diseases. Many serious epidemics were caused by the consumption of such products before this fact was clearly recognized.

However, this became less common as milk sanitation has improved and pasteurization is being more widely practised.

The disease organisms present in milk may be derived from:

- Diseased animals
- Persons collecting and handling milk

Thus the danger is due to the inoculum and not to the growth of organisms in the milk. The health of animal is an important factor.

Several diseases of cattle including staphylococcal and streptococcal infections, tuberculosis, brucellosis, salmonellosis, Q fever and Foot and mouth disease may be transmitted to man. The organisms causing these diseases may get into the milk either directly from the udder, or indirectly from infected body discharges, which may drop, splash, or be blown into the milk.

Some of the important diseases of human origin that have been transmitted by milk are:

- Typhoid fever
- Diphtheria,
- Scarlet fever,
- Dysentery
- Septic sore throat
- Poliomyelitis.

It is also possible for humans to infect animals. For example, mastitis may be caused by a variety of organisms, including Staphylococcus aureus. The infecting organism, in s9me cases, has been traced to humans.

CHARACTERISTICS OF MICROORGANISMS

MICROORGANISMS AND THEIR ROLE

The activated sludge process is a wastewater treatment method in which the carbonaceous organic matter of wastewater provides an energy source for the production of new cells for a mixed population of microorganisms in an aquatic aerobic environment. The microbes convert carbon into cell tissue and oxidized end products that include carbon dioxide and water. In addition, a limited number of microorganisms may exist in activated sludge that obtain energy by oxidizing ammonia nitrogen to nitrate nitrogen in the process known as nitrification.

Bacteria constitute the majority of microorganisms present in activated sludge. Bacteria that require organic compounds for their supply of carbon and energy (heterotrophic bacteria) predominate, whereas bacteria that use inorganic compounds for cell growth (autotrophic bacteria) occur in proportion to concentrations of carbon and nitrogen. Both aerobic and anaerobic bacteria may exist in the activated sludge, but the preponderance of species are facultative, able to live in either the presence of or lack of dissolved oxygen.

Fungi, rotifers, and protozoans are also residents of activated sludge. The latter microorganisms are represented largely by ciliated species, but flagellated protozoans and amoebae may also be present. Protozoans serve as indicators of the activated sluge condition, and ciliated species are instrumental in removing Escherichia coli from sewage. Additionally, viruses of human origin may be found in raw sewage influent, but a large percentage appear to be removed by the activated-sludge process.

The success of the activated-sludge process is dependent upon establishing a mixed community of microorganisms that will remove and consume organic waste material, that will aggregate and adhere in a process known as bioflocculation, and that will settle in such a manner as to produce a concentrated sludge (return activated sludge, or RAS) for recycling. Any of several types of activated sludge solids separations problems indicate an imbalance in the biological component of this process. In the ideal"healthy"system, filamentous organisms grow within a floc (a large aggregate of adherent, or floc-forming, microorganisms, such as bacteria) and give it strength, with few filaments protruding out into the surrounding bulk solution. In such a system, there is no interference with the compaction and settling rates of the activated sludge prior to its recycling.

INTRODUCTION

The activated-sludge process is a biological method of wastewater treatment that is performed by a variable and mixed community of microorganisms in an aerobic aquatic environment. These microorganisms derive energy from carbonaceous organic matter in aerated wastewater for the production of new cells in a process known as synthesis, while simultaneously releasing energy through the conversion of this organic matter into compounds that contain lower energy, such as carbon dioxide and water, in a process called respiration. As well. a variable number of microorganisms in the system obtain energy by converting ammonia nitrogen to nitrate nitrogen in a process termed nitrification. This consortium of microorganisms, the biological component of the process, is known collectively as activated sludge.

The overall goal of the activated-sludge process is to remove substances that have a demand for oxygen from the system. This is accomplished by the metabolic reactions (synthesis-respiration and nitrificaction) of the microorganisms, the separation and settling of activated-sludge solids to create an acceptable quality of secondary wastewater effluent, and the collection and recycling of microorganisms back into the system or removal of excess microorganisms from the system.

Physical Components Of Sludge Process

Before beginning a discussion of the biological component of the system, an overview of the physical components that comprise the activated-sludge process would seem to be in order. This will help the reader gain a better understanding of the environment in which a mixed community of microorganisms metabolizes organic (or in some cases inorganic) matter, settles to form a thickened sludge, and is recycled back into or removed from the system.

According to Activated Sludge, Manual of Practice #9 (Water Environment Association, 1987), the activated-sludge process contains five essential interrelated equipment components. The first is an aeration tank or tanks in which air or oxygen is introduced into the system to create an aerobic environment that meets the needs of the biological community and that keeps the activated sludge properly mixed. At least seven modifications in the shape and number of tanks exist to produce variations in the pattern of flow.

Second, an aeration source is required to ensure that adequate oxygen is fed into the tank(s) and that the appropriate mixing takes place. This source may be provided by pure oxygen, compressed air or mechanical aeration. Just as there are modifications in the shape and number of aeration tanks that can be used in the activated-sludge process, different equipment systems exist to deliver air or oxygen into aeration tanks.

SOLIDS SEPARATION

Microorganisms play a second important role in the activated-sludge process beyond removing carbonaceous organic material from and nitrifying ammonia in secondary influent wastewater. This is the process of solids separation in which activated-sludge solids separate by flocculation and gravity sedimentation from treated wastewater in secondary clarifers. The goal of this process is to create a secondary effluent low in suspended solids in the upper portion of a clarifier and a thickened activated sludge composed of flocs in the bottom portion of a clarifier that will be recycled back into the system as return activated sludge (RAS).

Activated sludge flocs, agglomerations of particles that may reach sizes of more than 1mm, are composed of the biological component discussed in the previous section and a nonbiological component. According to Jenkins et al. genera of heterotrophic bacteria, including Achromobacter, Alcaligenes,Arthrobacter, Citromonas, Flavobacterium, Pseudomonas, and Zoogloea, appear to be the primary floc-forming microorganisms. The nonbiological component of activated sludge flocs are organic and inorganic particles and extracellular microbial polymers that are generally made up largely of carbohydrates (polysaccharide [complex carbohydrate] and glycoprotein [a conjugated protein in which the nonprotein group is a carbohydrate] fibers).

Two levels of structure exist in activated-sludge flocs: microstructure and macrostructure. Microbial aggregation, adhesion, and bioflocculation are the basis of the microstructure. Although the mechanism of bioflocculation is not well understood, it is felt to be the result of bridging between extracellular microbial polymers functioning as polyelectrolytes (a substance of high molecular weight, such as a protein, that is an ionic conductor).

MICROORGANISMS IN NATURE

DEFINITION

The Theoretical Biology Group of the University of Bonn, Germany, visited the Wessex Institute and gave an overview of microbial survival strategies involving competition, cooperation, and communication, explaining the importance of spatial dynamics for these processes in his talk on "Survival strategies of microorganisms in nature".

Kinetic theory of optimal pathway length assumes that the total concentration of enzymes as well as metabolites in a cell is minimized because of the costs of enzyme synthesis and metabolite toxicity. Extending a given pathway by an extra step while constraining total concentration of enzymes and metabolites would lead to a reduction in the concentration of the already existing enzymes and intermediates. Assuming linear kinetics, where reaction rates are proportional to enzyme and metabolite concentrations, the rate of substrate consumption of a pathway decreases with the square of the pathway length, while the ATP yield of the pathway increases with its length, typically resulting in an optimal length in terms of the rate of ATP production or growth rate. The predictions of kinetic theory are in full agreement with observed lengths of pathways in nitrification, methane oxidation, aerobic and anaerobic sugar degradation, etc. However, a longer pathway would have a higher ATP or growth yield. This is the reason for the observed trade-off between growth rate and yield.

This trade-off has consequences for fitness, as can be seen by competing two growth strategies based on this trade-off. The strategy of fast growth at low yield, or low efficiency, (HighRate) is due to short pathways, whereas the strategy of slow growth at high yield (HighYield) is due to long pathways. In well-mixed systems, global competition for resources allows the HighRate strategy to win because it has the higher growth rate at any resource concentration. However, in spatially structured systems such as biofilms, the HighYield strategy has a higher fitness when the diffusional influx of substrate into the biofilm limits growth, because under such conditions, the higher efficiency of resource exploitation pays off, provided the neighbours cooperate in using the resource economically. This cooperation among neighbours can evolve because microbes attached to surfaces typically grow and multiply forming clusters of relatives; the offspring from a single ancestor stays together.

Microbes not only cooperate, they often do so only if they are not alone, and they use cell-to-cell communication to find out whether they are alone or in a group. They do this by producing a diffusible extracellular signal that switches on gene expression if a threshold concentration of the signal has been reached. However, the concentration of the signal not only depends on the density of signal producing cells, but also on diffusional constraints in the physical environment and on the degree of clustering of the cells. This clustering is essential for communication in complex natural environments, since it prevents interference from the multitude of other signalling species (practical problems) and exploitation by cheaters, cells that save costs because they do not produce signals or because they do not respond to the signal but nevertheless benefit from the signal producing and responding other cells (evolutionary problems).

A new, non-obligate system of bacterial predation on bacteria in soil was discovered. The predators comprising this system were a genetically unstable strain of Streptomyces venezuelae; a new kind of budding bacterium (now designated as Ensifer adhaerens, gen. n., sp. n.); and a myxobacterium. The natural population of these organisms in soil is attracted to and rapidly destroys (lyses) added Micrococcus luteus cells. There is also attack of predator on predator in this system. The transitory myceloid stage that occurs in the pleomorphic growth cycle of various Arthrobacter species was expanded (by manganese removal) so that it became the main form of growth. This stage was found to be more sensitive to bacteriophage attack than were the other growth stages. A new procedure for isolating Arthrobacter sp. bacteriophage from soil was developed for these studies. It was found that the encystment of Azotobacter vinelandii does not require prior intracellular production of poly-Beta-hydroxybutyrate, and that it is triggered by hypoxanthine.

Chapter 3

Water and Waste Water

HYDROLOGICAL CYCLE

DEFINITION

The sun's heat provides energy to evaporate water from the earth's surface (oceans, lakes, etc.). Plants also lose water to the air - this is called transpiration. The water vapour eventually condenses, forming tiny droplets in clouds.

When the clouds meet cool air over land, precipitation (rain, sleet, or snow) is triggered, and water returns to the land (or sea). Some of the precipitation soaks into the ground. Some of the underground water is trapped between rock or clay layers - this is called groundwater. But most of the water flows downhill as runoff (above ground or underground), eventually returning to the seas as slightly salty water.

This Information page provides an understanding of the hydrological cycle. It describes the principal stages of the cycle, with a brief description of each stage. A diagram gives a clear visual explanation. The links between the hydrological cycle and the duties of a water utility to supply clean water and dispose of dirty water are also explained.

EARTH'S WATER

Water is the most widespread substance to be found in the natural environment and it is the source of all life on earth. Water covers 70% of the earth's surface but it is difficult to comprehend the total amount of water when we only see a small portion of it. The distribution of water throughout the earth is not uniform. Some places have far more rainfall than others. There are many reasons for this, such as where the mountains are and where the prevailing winds blow. This rainfall distribution partly explains the differences in vegetation and why some areas are deserts and some are rainforests.

Water exists in three states: *liquid, solid* and invisible *vapour*. It forms the oceans, seas, lakes, rivers and the underground waters found in the top layers of the earth's crust and soil cover. In a solid state, it exists as ice and snow cover in polar and alpine regions. A certain amount of water is contained in the air as water vapour, water droplets and ice crystals, as well as in the biosphere. Huge amounts of water are bound up in the composition of the different minerals of the earth's crust and core.

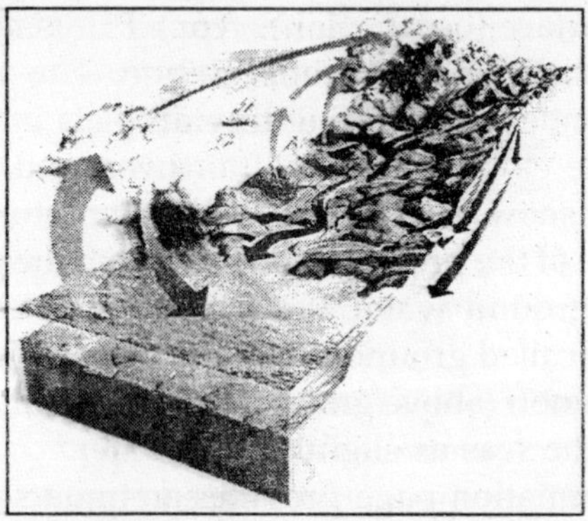

To assess the total water storage on the earth reliably is a complicated problem because water is so very dynamic. It is in permanent motion, constantly changing from liquid to solid or gaseous phase, and back again. The quantity of water found in the hydrosphere is the usual way of estimating the earth's water.

This is all the free water existing in liquid, solid or gaseous state in the atmosphere, on the Earth's surface and in the crust down to a depth of 2000 metres. Current estimates are that the earth's *hydrosphere* contains a huge amount of water - about 1386 million cubic kilometres. However, 97.5% of this amount exists as saline waters and only 2.5% as fresh water.

The greatest portion of the fresh water (68.7%) is in the form of ice and permanent snow cover in the Antarctic, the Arctic and in the mountainous regions. 29.9% exists as fresh *groundwaters*. Only 0.26% of the total amount of fresh water on the earth is concentrated in lakes, reservoirs and river system, where it is most easily accessible for our economic needs and absolutely vital for water ecosystems.

WORKING STAGES OF HYDROLOGICAL CYCLE

Evaporation

Water is transferred from the surface to the atmosphere through *evaporation*, the process by which water changes from a liquid to a gas. The sun's heat provides energy to evaporate water from the earth's surface. Land, lakes, rivers and oceans send up a steady stream of water vapour and plants also lose water to the air (transpiration). Approximately 80% of all evaporation is from the oceans, with the remaining 20% coming from inland water and vegetation.

Transport

The movement of water through the atmosphere, specifically from over the oceans to over land, is called *transport*. Some of the earth's moisture transport is visible as clouds, which themselves consist of ice crystals and/or tiny water droplets. Clouds are propelled from one place to another by either the jet stream, surface-based circulations like land and sea breezes or other mechanisms. However, a typical cloud 1 km thick contains only enough water for a millimetre of rainfall, whereas the amount of moisture in the atmosphere is usually 10-50 times greater than this.

Most water is transported in the form of water vapour, which is actually the third most abundant gas in the atmosphere. Water vapour may be invisible to us, but not to satellites which are capable of collecting data about

Condensation

The transported water vapour eventually *condenses*, forming tiny droplets in clouds.

Precipitation

The primary mechanism for transporting water from the atmosphere to the surface of the earth is *precipitation.* When the clouds meet cool air over land, precipitation, in the form of rain, sleet or snow, is triggered and water returns to the land (or sea). A proportion of atmospheric precipitation evaporates.

Groundwater

Some of the precipitation soaks into the ground and this is the main source of the formation of the waters found on land - rivers, lakes, groundwater and glaciers.

Some of the underground water is trapped between rock or clay layers - this is called *groundwater*. Water that infiltrates the soil flows downward until it encounters impermeable rock and then travels laterally. The locations where water moves laterally are called 'aquifers'. Groundwater returns to the surface through these *aquifers,* which empty into lakes, rivers and the oceans.

Under special circumstances, groundwater can even flow upward in artesian wells. The flow of groundwater is much slower than run-off with speeds usually measured in centimetres per day, metres per year or even centimetres per year.

Run-off

Most of the water which returns to land flows downhill as run-off. Some of it penetrates and charges groundwater while the rest, as river flow, returns to the oceans where it evaporates. As the amount of groundwater increases or decreases, the water table rises or falls accordingly. When the entire area below the ground is saturated, flooding occurs because all subsequent precipitation is forced to remain on the surface.

Different surfaces hold different amounts of water and absorb water at different rates. As a surface becomes less permeable, an increasing amount of water remains on the surface, creating a greater potential for flooding. Flooding is very common during winter and early spring because frozen ground has no permeability, causing most rainwater and meltwater to become run-off.

Water Balance

A considerable portion of river flow does not reach the ocean, having evaporated those areas with no natural surface run-off channels. On the other hand, some groundwater bypasses river systems altogether and goes directly to the ocean or evaporates.

Every year, the turnover of water on Earth involves 577,000 km^3 of water. This is water that evaporates from the ocean surface (502,800 km^3) and from land (74,200 km^3). The same amount of water falls as atmospheric precipitation, 458,000 km^3 on the ocean and 119,000 km^3 on land. The difference between precipitation and evaporation from the land surface (119,000?–?74,200 = 44,800 km^3/year) represents the total run-off of the Earth's rivers (42,700 km^3/year) and direct groundwater run-off.

Surface Waters

Rain water searches for the quickest route to the sea and flows into rivers, streams, lakes and underground stores. The water in the surface waters is clean enough to support a variety of wildlife. Although it is clean, it is not safe to drink and needs to be treated in a water treatment works to remove any substances that may be harmful.

OCEAN AND ATMOSPHERE

Water moves constantly between these reservoirs through the processes of evaporation, condensation and precipitation, surface and underground flow, and others. The driving force for the hydrologic cycle is the sun, which provides the energy needed for evaporation just as the flame of a gas stove provides the energy necessary to boil water and create steam. Water changes from a liquid state to a gaseous state as it evaporates from the oceans, lakes, streams, and soil (see our Water: Properties and Behaviour module for a further explanation).

Because the oceans are the largest reservoir of liquid water, that is where most evaporation occurs. The amount of water vapour in the air varies widely over time and from place to place; we feel these variations as humidity.

The presence of water vapour in the atmosphere is one of the things that makes earth livable for us. In 1859, Irish naturalist John Tyndall began studying the thermal properties of the gasses in the earth's atmosphere. He found that some gasses, like carbon dioxide (CO_2) and water vapour, trap heat in the atmosphere (a property commonly called the greenhouse effect), while other gasses like nitrogen (N_2) and argon (Ar) allow heat to escape to space. The presence of water vapour in the atmosphere helps keep surface air temperatures on the earth range from about -40° C to 55° C. Temperatures on planets without water vapour in the atmosphere, like Mars, stay as low as -100° C.

Once water vapour is in the air, it circulates within the atmosphere. When an air package rises and cools, the water vapour condenses back to liquid water around particulates like dust, called condensation nuclei. Initially these condensation droplets are much smaller than raindrops and are not heavy enough to fall as precipitation. These tiny water droplets create clouds. As the droplets continue to circulate within the clouds, they collide and form larger droplets, which eventually become heavy enough to fall as rain, snow, or hail. Though the amount of precipitation varies widely over the surface of the earth, evaporation and precipitation are globally balanced. In other words, if evaporation increases, precipitation also increases; rising global temperature is one factor that can cause a worldwide increase in evaporation from the world's oceans, leading to higher overall precipitation.

Since oceans cover around 70% of the earth's surface, most precipitation falls right back into the ocean and the cycle begins again. A portion of precipitation falls on land, however, and it takes one of several paths through the hydrologic cycle. Some water is taken up by soil and plants, some runs off into streams and lakes, some percolates into the groundwater reservoir, some falls on glaciers and accumulates as glacial ice.

HUMANS AND HYDROLOGIC CYCLE

The properties of water and the hydrologic cycle are largely responsible for the circulation patterns we see in the atmosphere and the oceans on the earth. Atmospheric and oceanic circulation are two of the major factors that determine the distribution of climatic zones over the earth. Changes in the cycle or circulation can result in major climatic shifts.

For example, if average global temperatures continue to increase as they have in recent decades, water that is currently trapped as ice in the polar ice sheets will melt, causing a rise in sea level. Water also expands as it gets warmer, further exacerbating sea level rise. Many heavily populated coastal areas like New Orleans, Miami, and Bangladesh will be inundated by a mere 1 meter increase in sea level. Additionally, the acceleration of the hydrologic cycle (higher temperatures mean more evaporation and thus more precipitation) may result in more severe weather and extreme conditions. Some scientists believe that the increased frequency and severity of El Niño events in recent decades is due to the acceleration of the hydrologic cycle induced by global warming.

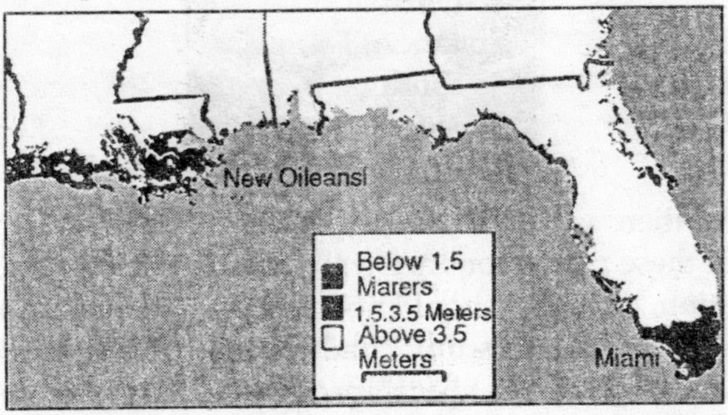

Even more immediately, the finitude of earth's fresh water resources is becoming more and more apparent. Groundwater can take thousands or millions of years to recharge naturally, and we are using these resources far faster than they are being replenished. The water table in the Ogallala Aquifer, which underlies 175,000 square miles of the US from Texas to South Dakota, is dropping at a rate of 10-60 cm per year due to extraction to irrigate the nation's bread basket. Surface waters around the world are largely contaminated by human and animal waste, most noticeably in countries like India and China, where untreated rivers provide the drinking and washing water for nearly 2 billion people.

Although legislation like the Clean Water Act in the US and water conservation practices such as the use of low-flow toilets and showerheads in parts of the world has begun to address these issues, the problems will only grow as world population increases. Every spring and well, every river and sea does indeed flow from the same source, and changes affect not just one river or lake, but the whole hydrologic cycle.

STRUCTURE OF WATER MOLECULE

A chemist's view of the world is not as narrow as one might think! Yes, we start with the atom, and then go on to the rules governing the kinds of structural units that can be made from them. We are taught early on to predict the properties of bulk matter from these geometric arrangements.

And then we come to H_2O, and are shocked to find that many of these predictions are way off, and that water (and by implication, life itself) should not even exist on our planet! But we soon learn that this tiny combination of three nuclei and ten electrons possesses special properties that make it unique among the more than 15 million chemical species we presently know. When we stop to ponder the consequences of this, chemistry moves from being an arcane science to a voyage of wonder and pleasure as we learn to relate the microscopic world of the atom to the greater world in which we all live.

MOLECULE OF WATER

A molecule is an aggregation of atomic nuclei and

electrons that is sufficiently stable to possess observable properties- and there are few molecules that are more stable and difficult to decompose than H_2O. In water, each hydrogen nucleus is bound to the central oxygen atom by a pair of electrons that are shared between them; chemists call this shared electron pair a covalent chemical bond.

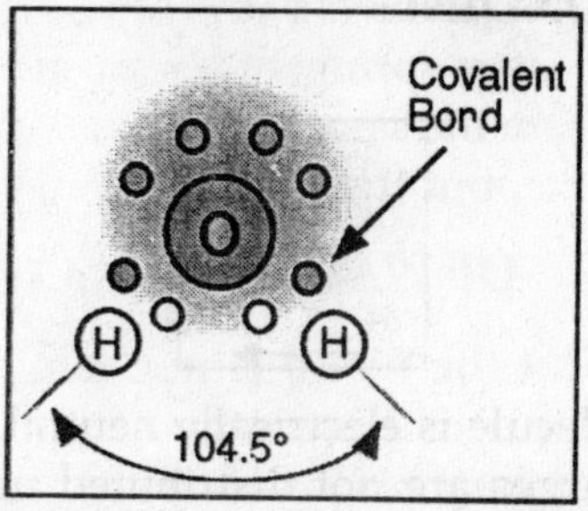

In H_2O, only two of the six outer-shell electrons of oxygen are used for this purpose, leaving four electrons which are organized into two non-bonding pairs. The four electron pairs surrounding the oxygen tend to arrange themselves as far from each other as possible in order to minimize repulsions between these clouds of negative charge. This would ordinarily result in a tetrahedral geometry in which the angle between electron pairs (and therefore the H-O-H bond angle) is 109.5°. However, because the two non-bonding pairs remain closer to the oxygen atom, these exert a stronger repulsion against the two covalent bonding pairs, effectively pushing the two hydrogen atoms closer together. The result is a distorted tetrahedral arrangement in which the H-O-H angle is 104.5°.

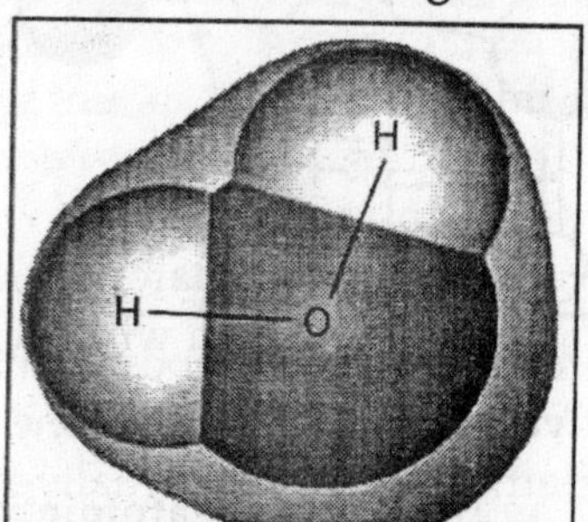

Because molecules are smaller than light waves, they

cannot be observed directly, and must be"visualized"by alternative means. This computer-generated image comes from calculations that model the electron distribution in the H_2O molecule. The outer envelope shows the effective"surface"of the molecule as defined by the extent of the cloud of negative electric charge created by the ten electrons.

HYDROGEN BONDING

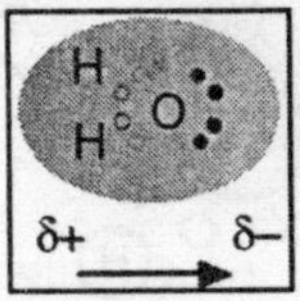

The H_2O molecule is electrically neutral, but the positive and negative charges are not distributed uniformly. This is illustrated by the gradation in Colour in the schematic diagram here. The electronic (negative) charge is concentrated at the oxygen end of the molecule, owing partly to the nonbonding electrons (solid blue circles), and to oxygen's high nuclear charge which exerts stronger attractions on the electrons. This charge displacement constitutes an *electric dipole,* represented by the arrow at the bottom; you can think of this dipole as the electrical "image" of a water molecule.

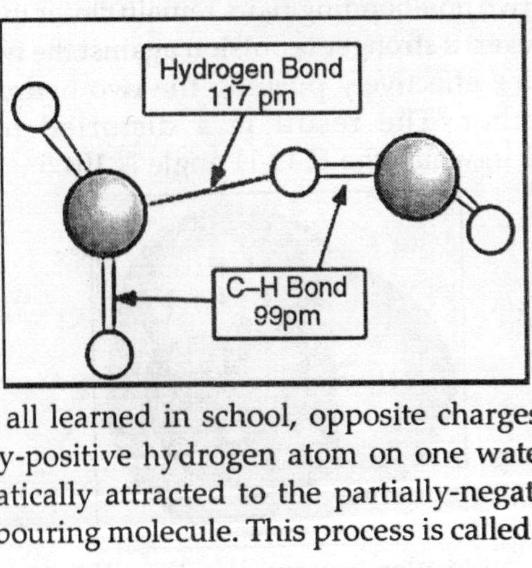

As we all learned in school, opposite charges attract, so the partially-positive hydrogen atom on one water molecule is electrostatically attracted to the partially-negative oxygen on a Neighbouring molecule. This process is called (somewhat misleadingly) *hydrogen bonding*. Notice that the hydrogen bond (shown by the dashed green line) is somewhat longer than the covalent O—H bond. This means that it is considerably*weaker*; it is so weak, in fact,that a given hydrogen bond cannot survive for more than a tiny fraction of a second.See here for much more about hydrogen bonding.

PROPERTIES OF WATER

Water has long been known to exhibit many physical properties that distinguish it from other small molecules of comparable mass. Chemists refer to these as the "anomalous" properties of water, but they are by no means mysterious; all are entirely predictable consequences of the way the size and nuclear charge of the oxygen atom conspire to distort the electronic charge clouds of the atoms of other elements when these are chemically bonded to the oxygen.

Water is one of the few known substances whose solid form is less dense than the liquid. The plot at the right shows how the volume of water varies with the temperature; the large increase (about 9%) on freezing shows why ice floats on water and why pipes burst when they freeze. The expansion between -4° and 0° is due to the formation of larger hydrogen-bonded aggregates. Above 4°, thermal expansion sets in as vibrations of the O—H bonds becomes more Vigourous, tending to shove the molecules farther apart.

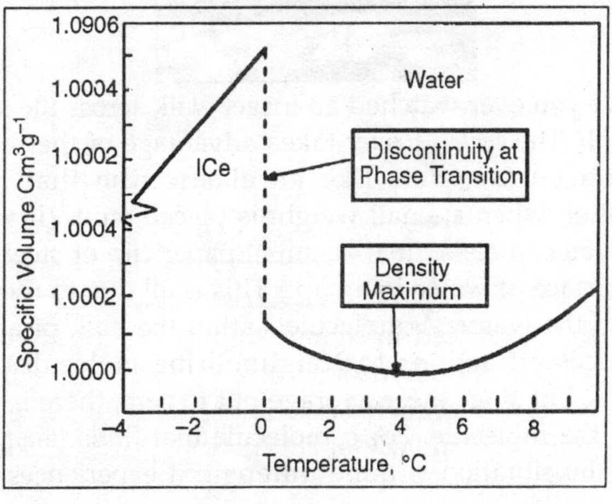

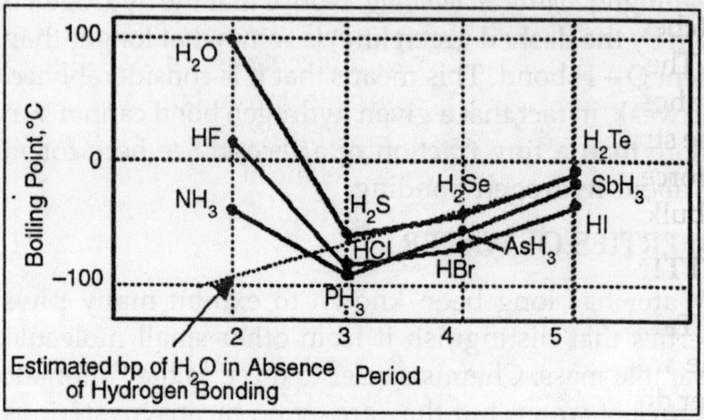

The other widely-cited anomalous property of water is its high boiling point. As this graph shows, a molecule as light as H_2O "should" boil at around –90°C; that is, it would exist in the world as a gas rather than a liquid if H-bonding were not present. Notice that H-bonding is also observed with fluorine and nitrogen.

SURFACE TENSION AND WETTING

Have you ever watched an insect walk across the surface of a pond? The *water strider* takes advantage of the fact that the water surface acts like an elastic film that resists deformation when a small weight is placed on it. (If you are careful, you can also "float" a small paper clip or steel staple on the surface of water in a cup.) This is all due to the*surface tension* of the water. A molecule within the bulk of a liquid experiences attractions to Neighbouring molecules in all directions, but since these average out to zero, there is no net force on the molecule. For a molecule that finds itself *at* the surface, the situation is quite different; it experiences forces only sideways and downward, and this is what creates the stretched-membrane effect.

The distinction between molecules located at the surface and those deep inside is especially prominent in H_2O, owing to the strong hydrogen-bonding forces. The difference between the forces experienced by a molecule at the surface and one in the bulk liquid gives rise to the liquid's *surface tension.*

WETTING

Take a plastic mixing bowl from your kitchen, and splash some water around in it. You will probably observe that the water does not cover the inside surface uniformly, but remains dispersed into drops. The same effect is seen on a dirty windshield; turning on the wipers simply breaks hundreds of drops into thousands. By contrast, water poured over a clean glass surface will *wet*it, leaving a uniform film.

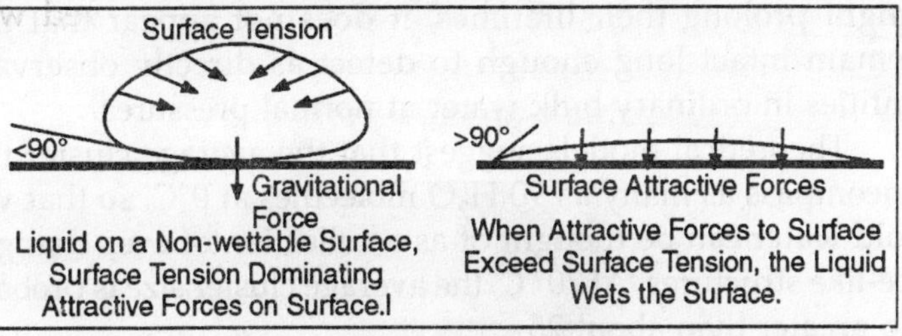

Liquid on a Non-wettable Surface, Surface Tension Dominating Attractive Forces on Surface.I

When Attractive Forces to Surface Exceed Surface Tension, the Liquid Wets the Surface.

When a liquid is in contact with a solid surface, its Behaviour depends on the relative magnitudes of the surface tension forces and the attractive forces between the molecules of the liquid and of those comprising the surface. If an H_2O molecule is more strongly attracted to its own kind, then surface tension will dominate, increasing the curvature of the interface.

This is what happens at the interface between water and a hydrophobic surface such as a plastic mixing bowl or a windshield coated with oily material. A clean glass surface, by contrast, has -OH groups sticking out of it which readily attach to water molecules through hydrogen bonding: this causes the water to spread out evenly over the surface, or to wet it. A liquid will wet a surface if the angle at which it makes contact with the surface is more than 90°. The value of this*contact angle* can be predicted from the properties of the liquid and solid separately.

If we want water to wet a surface that is not ordinarily wettable, we add a*detergent* to the water to reduce its surface tension. A detergent is a special kind of molecule in which one end is attracted to H_2O molecules but the other end is not; the latter ends stick out above the surface and repel each other, cancelling out the surface tension forces due to the water molecules alone.

STRUCTURES AND WATER CLUSTERS

It is quite likely that over very small volumes, localized $(H_2O)_n$ polymeric clusters may have a fleeting existence, and many theoretical calculations have been made showing that some combinations are more stable than others. While this might prolong their lifetimes, it does not appear that they remain intact long enough to detect as directly observable entities in ordinary bulk water at normal pressures.

Theoretical models suggest that the average cluster may encompass as many as 90 H_2O molecules at 0°C, so that very cold water can be thought of as a collection of ever-changing ice-like structures. At 70° C, the average cluster size is probably no greater than about 25.

It must be emphasized that no stable clustered unit or arrangement has ever been isolated or identified in pure bulk liquid water. A 2006 reportsuggests that a simple tetrahedral arrangement is the only long-range structure that persists at time scales of a picosecond or beyond. But for an interesting (and somewhat controversial) alternative view, see this article byRustum Roy. And a 2007 study suggests that infrared radiation can stabilize clathrate-like clusters for up to several hours.

Water clusters are of considerable interest as models for the study of water and water surfaces, and many articles on them are published every year. Some notable work reported in 2004 extended our view of water to the femtosecond time scale. The principal finding was that 80 per cent of the water molecules are bound in chain-like fashion to only two other molecules at room temperature, thus supporting the prevailing view of a dynamically-changing, disordered water structure.

LIQUID AND SOLID WATER

Ice, like all solids, has a well-defined structure; each water molecule is surrounded by four Neighbouring H_2Os. two of these are hydrogen-bonded to the oxygen atom on the central H_2O molecule, and each of the two hydrogen atoms is similarly bonded to another Neighbouring H_2O.

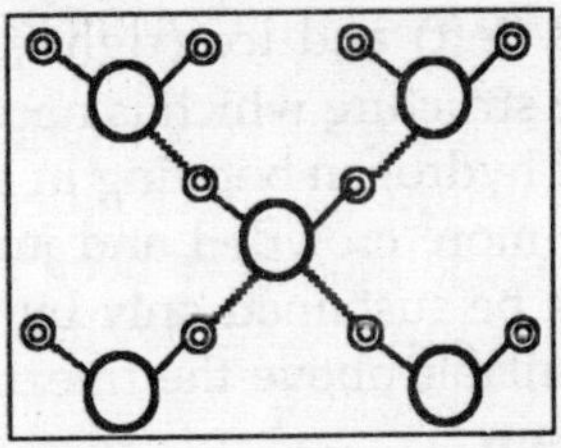

The hydrogen bonds are represented by the dashed lines in this 2-dimensional schematic diagram. In reality, the four bonds from each SO atom point toward the four corners of a tetrahedron Centreed on the O atom. This basic assembly repeats itself in three dimensions to build the ice crystal.

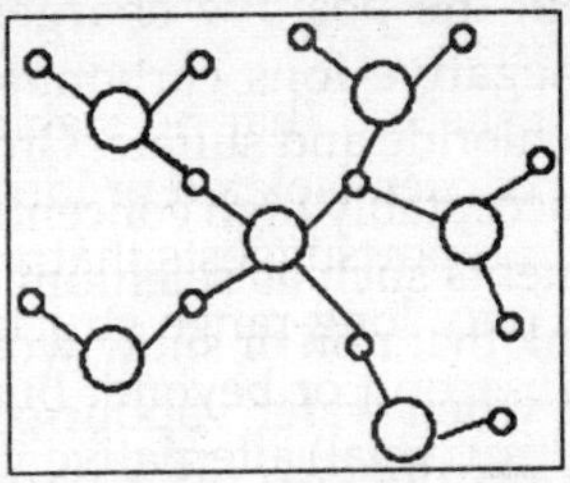

When ice melts to form liquid water, the uniform three-dimensional tetrahedral organization of the solid breaks down as thermal motions disrupt, distort, and occasionally break hydrogen bonds. The methods used to determine the positions of molecules in a solid do not work with liquids, so there is no unambiguous way of determining the detailed structure of water. The illustration here is probably typical of the arrangement of Neighbours around any particular H2O molecule, but very little is known about the extent to which an arrangement like this gets propagated to more distant molecules.

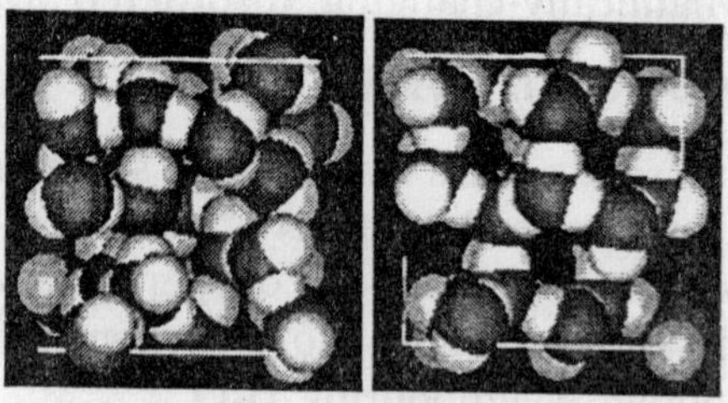

Here are three-dimensional views of a typical local structure of water (left) and ice (right.) Notice the greater openness of the ice structure which is necessary to ensure the strongest degree of hydrogen bonding in a uniform, extended crystal lattice. The more crowded and jumbled arrangement in liquid water can be sustained only by the greater amount thermal energy available above the freezing point.

DRINKING WATER

Our ordinary drinking water, by contrast, is never chemically pure, especially if it has been in contact with sediments. Groundwaters (from springs or wells) always contain ions of calcium and magnesium, and often iron and manganese as well; the positive charges of these ions are balanced by the negative ions carbonate/bicarbonate, and occasionally some chloride and sulfate. Groundwaters in some regions contain unacceptably high concentrations of naturally-occuring toxic elements such as selenium and arsenic.

One might think that rain or snow would be exempt from contamination, but when water vapour condenses out of the atmosphere it always does so on a particle of dust which releases substances into the water, and even the purest air contains carbon dioxide which dissolves to form carbonic acid. Except in highly polluted atmospheres, the impurities picked up by snow and rain are too minute to be of concern.

Various governments have established upper limits on the amounts of contaminants allowable in drinking water; the best known of these are theU.S. EPA Drinking Water Standards.

WATER IN OUR BODIES

About two-thirds of the weight of an adult human consists of water. About two-thirds of this water is located within cells, while the remaining third consists of extracellular water, mostly in the blood plasma and in the interstitial fluid that bathes the cells.

This water, amounting to about five per cent of body weight (about 5 L in the adult), serves as a supporting fluid for the blood cells and acts as a means of transporting chemicals between cells and the external environment. It is basically a 0.15M solution of salt (NaCl) containing smaller amounts of other electrolytes, the most important of which are bicarbonate (HCO_3^-) and protein anions.

The water content of our bodies is tightly controlled in respect to both total volume and its content of dissolved substances, particulary ions.

Drinking constitutes only one source of our water; many foods, especially those containing cells (fruits, vegetables, meats) are an important secondary source. In addition, a considerable amount of water (350-400 mL/day) is produced *metabolically* — that is, from the oxidation of glucose derived from foods.

The quantity of water exchanged within various parts of our bodies is surprisingly large. The kidneys process about 180 L/day, returning most of the water to the blood stream. Lymph flow amounts to 1-2.5 L/day, and turnover of fluids in the bowel to 8-9 L/day. These figures are dwarfed by the 80,000 L/day of water that diffuses in both directions through capillary walls.

WATER AS A SOLVENT

The partial charge that develops across the water molecule helps make it an excellent solvent. Water dissolves many substances by surrounding charged particles and "pulling" them into solution.

For example, common table salt, sodium chloride, is an ionic substance that contains alternating sodium and chlorine ions.

Fig. Sodium chloride contains Na^+ and Cl^- ions.

When table salt is added to water, the partial charges on the water molecule are attracted to the Na^+ and Cl^- ions. The water molecules work their way into the crystal structure and between the individual ions, surrounding them and slowly dissolving the salt. The water molecules will actually line up differently depending on which ions are being pulled into solution. The negative oxygen ends of water molecules will surround the positive sodium ions; the positive hydrogen ends will surround the negative chlorine ions.

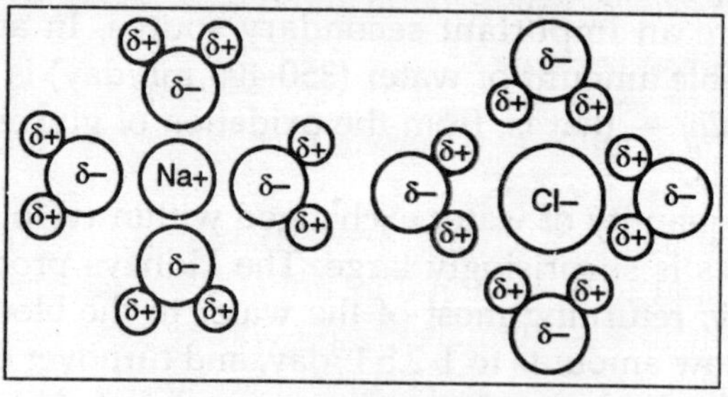

Fig: Salt Dissolving in Water

In a similar fashion, any substance that carries a net electrical charge, including both ionic compounds and polar covalent molecules (those that have a dipole), can dissolve in water. This idea also explains why some substances do not dissolve in water. Oil, for example, is a nonpolar molecule. Because there is no net electrical charge across an oil molecule, it is not attracted to water molecules and therefore does not dissolve in water.

The human body is 66% water by weight, according to Hill and Kolb. Water is the universal solvent for life, referred to by Nobel Laureate A. Szent-Gyorgy as "the matrix of life".That water serves as the solvent for sodium chloride (salt) and other substances so that the fluids of our bodies are similar to sea water. This leads Hill and Kolb to refer jokingly to us as "walking bags of sea water".

Water serves to suspend the red blood cells to carry oxygen to the cells. It is the solvent for the electrolytes and nutrients needed by the cells, and also the solvent to carry waste material away from the cells. With water as the solvent, osmotic pressure acts to transport the needed water into cells. With cells bathed in the interstitial fluid, diffusion contributes to carrying needed molecules into the cells

PRINCIPLES OF CHEMICAL EQUILIBRIUM

Equilibrium. In this chapter the dynamic equilibrium established in chemical reactions is explored, and its similarity to phase equilibrium is emphasized. The chapter culminates in a relationship between two signposts of spontaneity: the standard free energy of reaction and the equilibrium constant.

SPECIFIC CONCEPTS

Dynamic equilibrium is a balance of opposing processes occurring at equal rates.

- Chemical reactions achieve dynamic equilibrium similar to phase equilibrium and characterized by an equilibrium constant, K_{eq}, analogous to vapour pressure.

  Equilibrium as a balance of opposing rates. This state of balance gives the appearance of stasis at the macroscopic level, despite the continuous and dynamic activity at the microscopic level. Phase equilibrium is dynamic equilibrium. We summarize here the characteristics of dynamic equilibrium:
- It occurs only in closed systems.
- At least 2 states of a system must be simultaneously present (coexist).
- It is established only over a finite time interval, required to establish the equality of opposing rates.
- Values of macroscopic observables remain constant

with time, but at the molecular level there is constant transit among the states present
- When subjected to a perturbation from outside, it responds to minimize the effect of the perturbation and returns to equilibrium, again over a finite time interval.
- It represents a compromise between the tendencies to minimize potential energy and maximize disorder.

In this, we extend the dynamic equilibrium concept to chemical reactions. We will see that, like a pure subscance in a closed container, a chemical reaction in a closed system reaches a state of dynamic equilibrium, a state in which conversion of reactants to products is balanced by conversion of products to reactants. At the macroscopic scale, there is no change in the amount of a particular reactant or product with time. In contrast, at the molecular level there is a continuous reshuffling of atoms.

CHEMICAL EQUILIBRIUM

We introduce chemical equilibrium with a simple reaction, the dimerization of nitrogen dioxide to dinitrogen tetroxide, in equation:

$2\ NO_2 \rightarrow N_2O_4$

Figure 12-1
Apparatus for Gas Phase Equilibrum

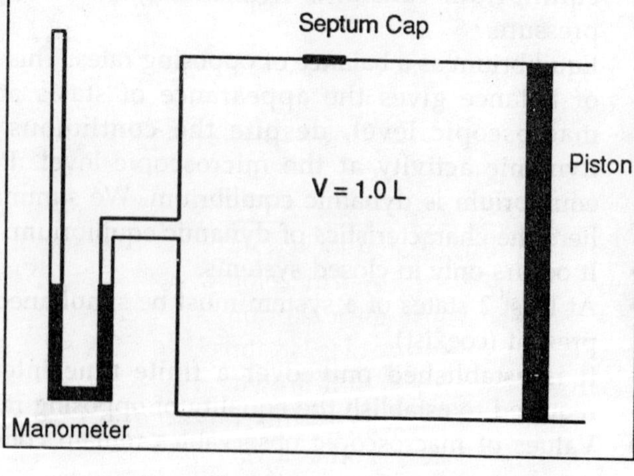

As we have done several times before, we will study the reaction by carrying out several thought experiments designed to reveal the Behaviour of the reaction under various conditions.

Throughout the following discussion, bear in mind that the experiments could be (and in fact have been) carried out in the Labouratory (although in a somewhat different manner than described). Above figure shows an initially evacuated container of volume 1.0 L, with an attached manometer. The manometer allows the total gas pressure in the container to be measured. The temperature of the container is 298 K. The container is fitted with a piston that can be freely moved or locked in place. The following experiments are sequentially performed.

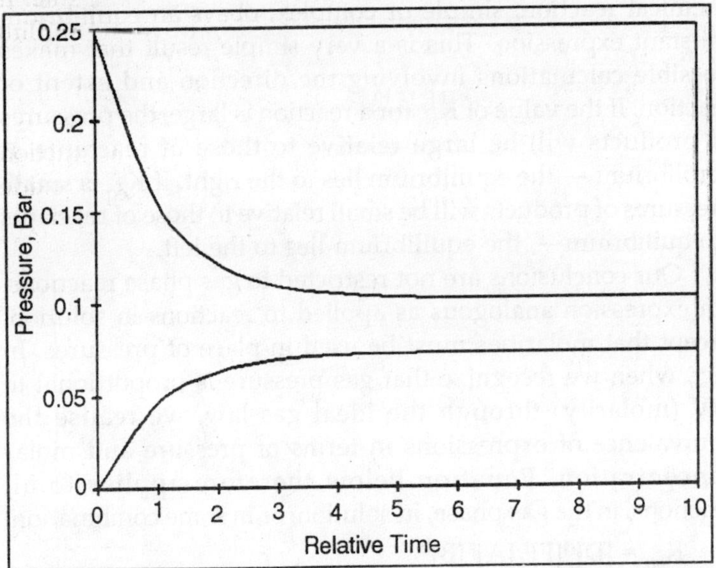

Expt 1. With the piston locked in place, 0.0460 g (0.0100 moles) of NO2 gas (red-brown in Colour, and toxic!) is injected into the container through a septum cap. The manometer quickly registers a pressure of 0.248 atm (188 mm of Hg). Then, over several minutes, the red-brown Colour fades and the pressure of the system falls, at first rapidly, then more and more slowly, until it levels off at a value of 0.175 atm.

Concept

The results of experiments on many different reactions over many years allow us to generalize as follows. For the generic reaction, in which the participating species are gases, the partial pressures at equilibrium obey the simple expressions, independent of the initial pressures of the gases.

$$a\,A + bB \rightarrow dD + fF$$

$$K_{eq} = P_D{}^d P_F{}^f / P_A{}^a P_B{}^b$$

In words, the equation says that if we divide the product of the pressures of the products, each raised to the power of its stoichiometric coefficient in the equation for the reaction, by a similar product of reactant pressures, the result is a constant, K_{eq}, that depends only on temperature. Every chemical reaction, simple or complex, obeys an equilibrium constant expression. This is a very simple result that makes possible calculations involving the direction and extent of reaction. If the value of K_{eq} for a reaction is large, the pressures of products will be large relative to those of reactants at equilibrium — the equilibrium lies to the right. If K_{eq} is small, pressures of products will be small relative to those of reactants at equilibrium — the equilibrium lies to the left.

Our conclusions are not restricted to gas phase reactions. An expression analogous as applied to reactions in solution, except that molarities must be used in place of pressures. In fact, when we recognize that gas pressure is proportional to n/V (molarity) through the ideal gas law, we realise the equivalence of expressions in terms of pressure and molar concentration. Equation below therefore applies to all reactions, in the gas phase, in solution, or in some combination:

$$K_{eq} = [D]^d[F]^f/[A]^a[B]^b$$

The brackets signify molarity. One caution is in order for gas phase reactions. The numerical value and the units of the equilibrium constant depend on whether pressures or molarities are used. If K_{eq} is specified in units of bar, pressures are used in calculations; if in moles/L, concentrations are used. If K_{eq} has no units, either pressure or molarity may be used.

LE CHATELIER'S PRINCIPLE

This principle applies as well to chemical equilibrium as to phase equilibrium. In fact, the principle is a qualitative manifestation of the quantitative characteristics of the equilibrium constant. Three stresses are particularly relevant to chemical equilibrium.

- Change in concentration (pressure) of a participant. For example, consider the reaction

$$N_2 + 3H_2 \rightarrow 2NH_3$$

Assume the reaction to be initially at equilibrium. Suppose that we add some NH_3 to the system. This imposes a stress on the system that will be relieved, according to Le Chatelier, by a shift that relieves the stress – i.e., uses up NH_3. The reaction shifts left until enough NH_3 is used and N_2 and H_2 produced to again make $Q = K_{eq}$. When the new equilibrium is reached, the pressures of all three gases will be larger than they were before the stress. Note that addition of NH_3 to the system does not change the value of K_{eq}. The same value applies before and after the stress.

Molecular View of Chemical Equilibrium

Consider now the general reaction in equations.

$A + B \rightarrow 2C$

$K_{eq} = P_C^2/P_AP_B$

Why do the stoichiometric coefficients appear as exponents in the expression for K_{eq}? Chemical equilibrium is like phase equilibrium; it results from a balance of opposing rates. In this case, the rates involved are those of the reaction in the forward and reverse directions:

At equilibrium, $R_f = R_r$

We assume now, and explore more fully later, that molecules must collide in order to react. This is certainly sensible, because it is not possible for atom rearrangements to occur unless the reacting atom aggregates are in close proximity. The rate of a chemical reaction is thus expected to be proportional to the number of collisions between molecules per unit time: rate of rxn proportional to #collisions/time

For the forward and reverse processes, we write,

$R_f = k_f$ (#collisions of A and B)

$R_r = k_r'$(#collisions of C and C)

The rate constants, k_f and k_r', are proportionality constants in the same way that k_E and k_C are. If there are N_A molecules of A and N_B of B in a container, a particular A molecule may collide with any of the N_B molecules of B. The total number of ways in which A and B may collide is thus $N_A \times N_B$.

Then,

$R_f = k_f N_A N_B$

N_C molecules of C can collide in a total of $N_C(N_C-1)/2$ ways. This arises because each molecule can collide with N_C-1 others. However, the simple product of N_C and N_C-1 double counts the collisions (i.e., it includes collisions of molecule x with y and of y with x; however, there can be only one collision between two particular molecules). We divide by 2 to eliminate double counting.

Finally, because N_C is very large, $N_C - 1 = N_C$. It follows that:

$R_r = k_r' N_C^2/2 = k_r N_C^2$

At chemical equilibrium, then,

$k_f N_A N_B = k_r N_C^2$

Rearranging gives,

$k_f/k_r = N_C^2/N_A N_B = K_{eq}$

Since pressure is proportional to number of molecules, this equation is readily translated to pressure terms, giving the equilibrium constant expression in equation. Notice that, as was true for phase equilibrium, the equilibrium constant is a ratio of rate constants for forward and reverse processes. Chemical equilibrium, like phase equilibrium, is dynamic rather than static. Although it appears at the macroscopic level that nothing is happening since concentrations remain constant, both forward and reverse reactions continue at equal rates at the molecular level.

FREE ENERGY AND EQUILIBRIUM

To this point we have encountered two quantities that indicate the extent to which a chemical reaction will proceed:

its spontaneity. These are K_{eq} and the standard free energy change of reaction, $DG_R{}^o$. We now explore the relationship between these quantities. The sign of $DG_R{}^o$ indicates the spontaneity or non-spontaneity of a process in which reactants in their standard states are converted to products in their standard states. The magnitude of $DG_R{}^o$ tells us how far the standard state is from the equilibrium state. A process for which $DG_R{}^o$ is negative is spontaneous under standard conditions, and will proceed left to right to reach equilibrium. At equilibrium, the reactants will be present at less than the standard state pressure (1 atm) or concentration (1 M), because they have been used up to some extent in achieving equilibrium. The products will be present at greater than the standard state pressure or concentration, because they have been formed in achieving equilibrium. In the equilibrium state, products will be present in relatively large amounts, reactants will be present in relatively small amounts, consistent with $K_{eq} > 1$.

Thus $DG_R{}^o < 0$ corresponds to $K_{eq} > 1$. Further, for a process with $DG_R{}^o = -50$ kJ, the standard state is much farther away from equilibrium than is the standard state for a process with $DG_R{}^o = -5$ kJ. The first reaction will have to proceed further to reach the equilibrium state than will the second reaction, and will thus have a larger K_{eq}. We can make very similar statements about processes for which $DG_R{}^o > 0$, except in reverse. The larger the magnitude of $DG_R{}^o$, the further the standard state is from the equilibrium position, and the further reaction will have to proceed to reach equilibrium from the standard state. For $DG_R{}^o > 0$, reaction must proceed from right to left toward equilibrium, decreasing the pressures or concentrations of products and increasing those of reactants. Thus $K_{eq} < 1$. The larger the magnitude of $DG_R{}^o$, the smaller K_{eq}. Finally, when $DG_R{}^o = 0$, the process is equally spontaneous in both forward and reverse directions; the standard state is the equilibrium state.

Supplement Applications
Consider the Reaction Below

$$2NO_2(g) \rightarrow N_2O_4(g) \quad K_{eq} = 7.0 \text{ atm}^{-1}$$

If you mix 0.12 atm NO_2 and 0.43 atm N_2O_4 in a closed bulb, will any reaction occur? In which direction? (you must provide convincing support for your answers.)

What will the partial pressures of NO_2 and N_2O_4 be once equilibrium is attained? Consider the reaction below:

$2NOBr(g) \rightarrow 2NO(g) + Br_2(g)$

Over a time interval between t_0 and t_1, the reaction is at equilibrium at some temperature, with constant partial pressures of reactants and products as At time t_1, the volume of the reaction system is instantly halved. Show on the graphs the immediate effect of the volume decrease on the partial pressures of species, and the manner in which these pressures change as equilibrium is reestablished. Your graphs should be as quantitatively correct as possible.

You are studying the equilibrium $P_4(g)$ $2P_2(g)$. In a previously evacuated vessel, you place a quantity of diatomic phosphorus gas. What is the value of the reaction quotient Q? Which direction will reaction proceed to achieve equilibrium?

You allow the system to come to equilibrium and then decide to tamper with it. In which direction would the equilibrium shift if you add $P_4(g)$? Increase the pressure of the system (at const T)? Increase the volume of the system (at const T)? Increase the T of the system (at const P)? What do you know about a system at equilibrium when you know that the value of the equilibrium constant is 0.34 at a particular temperature?

pH

INTRODUCTION

The water chemistry requirements for turtles are often underestimated. At the same time it needs to be acknowledged that they are not fish and hence are not susceptible to going belly up from a rapid pH change.

Some species do benefit from a more exacting regime, and it can be said that there are health benefits, even if indirect, from careful monitoring and adjustment of these water quality parameters. First of all I will define these terms as I think we all probably have some knowledge of what pH means but some detailed knowledge is required.

PER HYDROGEN (pH)

The pH of the water is a measure of the balance between the Hydrogen (H+) and Hydroxide (OH) ions in the water. I think most of us know that low pH is acidic and high pH is alkaline or basic. Hence a pH of 5 is slightly acidic water, a pH of 7 is neutral and a pH of 8 is alkaline water.

For the budding chemists the equation is as follows:

$$pH = -\log 10[H_3O^+]$$

All right enough of the chemistry but there is an important point here and any mathematicians will see it. The pH scale is a logarithmic scale. In other words a pH of 6.0 is 10 times more acidic than a pH of 7.0 and a pH of 5.0 is 100 times more acidic than a pH of 7.0. As a keeper you need to be aware that it is not a difference of 1 when you go from 6.0 to 7.0.

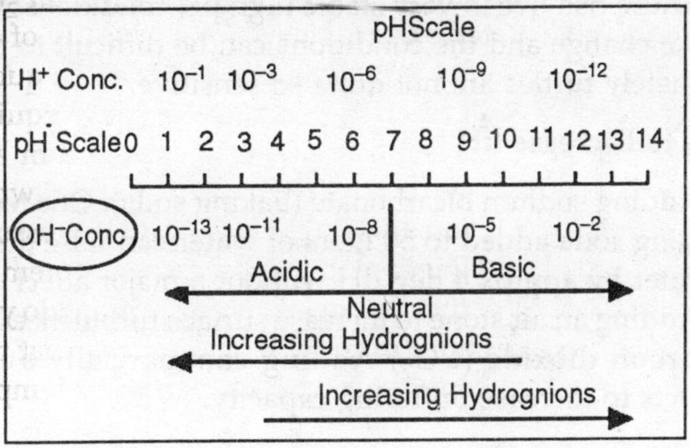

WAYS TO LOWER pH

Filtering water over peat Add bogwood to the tank Inject carbon dioxide CO_2 Use a commercial acid buffer Water changes with softened water or RO (Reverse Osmosis) water.

WAYS TO RAISE THE pH

Aerate the water, driving off the carbon dioxide (CO_2) Filter over coral or limestone Add rocks containing limestone to the tank or use a coral sand substrate Use a commercial alkaline buffer.

CARBONATE HARDNESS (kH)

This is an area where many people get confused. One of the reasons the term alkaline is avoided a bit and the term Basic is used for the pH scale is because this reading is measuring the alkalinity of the water. It is not the same as alkaline.

The alkalinity is a measurement of the waters buffering ability, or its ability to absorb and neutralise acid. Clearly the more alkalinity or the higher the Carbonate Hardness of the water the less likely you will incur pH swings in the water. It is therefore important to get this figure reasonably high to stabilise the water.

Exactly how high you want your dKH (degrees of KH) will depend on what pH you choose to use. To people keeping fish from the African Lakes this is the life and death of their fish. Those fish live in very stable high pH conditions, they do not like change and the conditions can be difficult to imitate. Fortunately turtles are not quite so sensitive.

Ways to Increase kH

Adding sodium bicarbonate (baking soda). One teaspoon of baking soda added to 50 liters of water can raise the kH of the water by approx 4 deg dH without a major affect on pH.

Adding an air stone to increase surface turbulence driving off carbon dioxide (CO_2) Adding commercially available products to increase buffering capacity.

Ways to lower kH

Injecting carbon dioxide (CO_2) Use reverse osmosis (RO) water. You can mix tap water with reverse osmosis water to achieve the desired kH. Adding commercially available products to decrease the buffering capacity.

Do not use distilled water as it has no dissolved salts and hence no buffering ability. Add a small amount of acid (eg uric acid) and it will shift the pH very rapidly. It is also highly osmotic and will react with the turtles renal system.

GENERAL HARDNESS (gH)

This is essentially a measurement of Magnesium and

Calcium ions in the water. Again it is measured in the German degrees of hardness scale or parts per million. This is what is generally meant by soft and hard water which are terms people should be familiar with. The table below shows comparisons between parts per million, the dH scale and the generalised concepts of soft and hard water.

Increasing gH

Adding limestone to the aquarium (this will also increase kH which in turn will increase pH) Adding calcium carbonate will raise gH and kH

Reducing gH

Adding peat moss to your filter Use commercially available water softening pillows or a water softener (this removes calcium and magnesium ions and replaces them with sodium ions. Many people feels that this is an unacceptable method of softening water as many fish that prefer soft water don't like sodium either.

TESTING THE WATER

This is actually relatively simple. The commercially available aquarium pH and Hardness test kits are certainly accurate enough. Follow the directions on whichever ones you buy and do the tests once per week to monitor the aquarium. Remember that when doing water changes you need to prepare the water so that it's the same as what you removed. So if you add salt you need to add more. If you are maintaining a high pH and your local water is soft and acidic you must correct this first.

OXIDATION

DEFINITION

A freshly-cut apple turns brown, a bicycle fender becomes rusty and a copperpenny suddenly turns green. What do all of these events have in common? They are all examples of a process called oxidation.

Oxidation is defined as the interaction between oxygen

molecules and all the different substances they may contact, from metal to living tissue.

Technically, however, with the discovery of electrons, oxidation came to be more precisely defined as the loss of at least one electron when two or more substances interact. Those substances may or may not include oxygen. (Incidentally, the opposite of oxidation is reduction - the addition of at least one electron when substances come into contact with each other.) Sometimes oxidation is not such a bad thing, as in the formation of super-durable anodized aluminum. Other times, oxidation can be destructive, such as the rusting of an automobile or the spoiling of fresh fruit.

We often used the words oxidation and rust interchangeably, but not all materials which interact with oxygen molecules actually disintegrate into rust. In the case of iron, the oxygen creates a slow burning process, which results in the brittle brown substance we call rust. When oxidation occurs in copper, on the other hand, the result is a greenish coating called copper oxide. The metal itself is not weakened by oxidation, but the surface develops a patina after years of exposure to air and water.

When it involves oxygen, the process of oxidation depends on the amount of oxygen present in the air and the nature of the material it touches. True oxidation happens on a molecular level - we only see the large-scale effects as the oxygen causes free radicals on the surface to break away. In the case of fresh fruit, the skin usually provides a barrier against oxidation. This is why most fruits and vegetables arrive in good condition at the grocery store. Once the skin has been broken, however, the individual cells come in direct contact with air and the oxygen molecules start burning them. The result is a form of rust we see as brownish spots or blemishes.

Oxidation can also be a problem for car owners, since the outermost layers of paint are constantly exposed to air and water. If the car's outer finish is not protected by a wax coating or polyurethane, the oxygen molecules in the air will eventually start interacting with the paint. As the oxygen burns up the free radicals contained in the paint, the finish becomes duller and duller. Restoration efforts may include removing several layers of affected paint and reapplying a new layer of protectant. This is why professional car detailers recommend at least one layer of wax or other protectant be used every time the car is washed.

The secret of preventing oxidation caused by oxygen is to provide a layer of protection between the exposed material and the air. This could mean a wax or polyurethane coating on a car, a layer of paint on metal objects or a quick spray of an anti-oxidant, like lemon juice, on exposed fruit. Destructive oxidation cannot occur if the oxygen cannot penetrate a surface to reach the free radicals it craves.

This is why *stainless steel* doesn't rust and ordinary steel does. The stainless steel has a thin coating of another metal which does not contain free radicals. Regular steel may be painted for protection against oxidation, but oxygen can still exploit any opening, no matter how small. This is why you may find a painted metal bicycle still damaged by rust.

OXIDATION AND REDUCTION

- Oxidation is gain of oxygen.
- Reduction is loss of oxygen.

For example, in the extraction of iron from its ore:

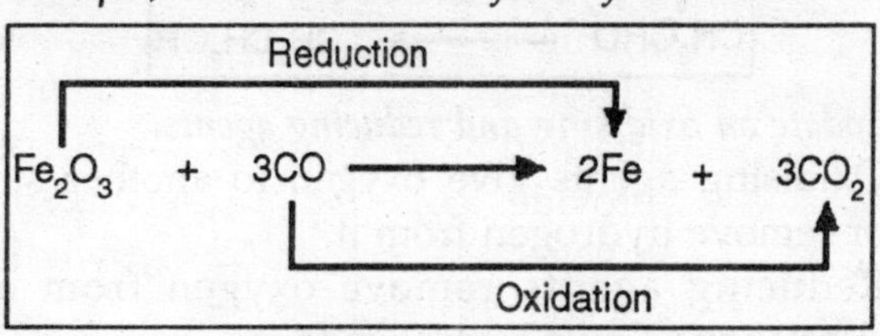

Because both *red*uction and *ox*idation are going on side-by-side, this is known as a *redox* reaction.

Oxidising and reducing agents

An oxidising agent is substance which oxidises something else. In the above example, the iron(III) oxide is the oxidising agent.

A reducing agent reduces something else. In the equation, the carbon monoxide is the reducing agent.

- Oxidising agents give oxygen to another substance.
- Reducing agents remove oxygen from another substance.

These are old definitions which aren't used very much nowadays. The most likely place you will come across them is in organic chemistry. Notice that these are exactly the opposite of the oxygen definitions. For example, ethanol can be oxidised to ethanal:

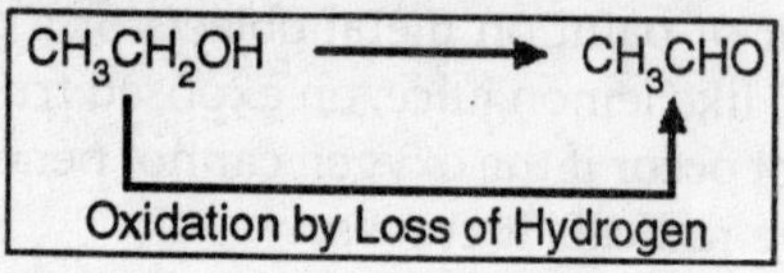

You would need to use an oxidising agent to remove the hydrogen from the ethanol. A commonly used oxidising agent is potassium dichromate(VI) solution acidified with dilute sulphuric acid.

Ethanal can also be reduced back to ethanol again by adding hydrogen to it. A possible reducing agent is sodium tetrahydridoborate, $NaBH_4$. Again the equation is too complicated to be worth bothering about at this point.

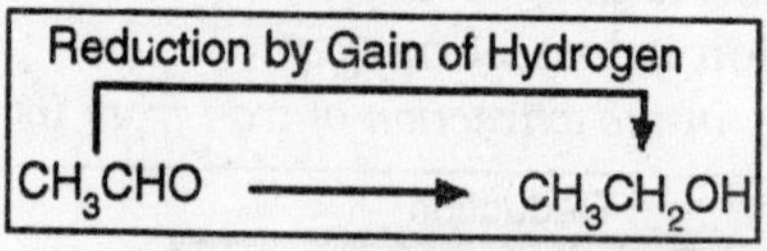

An update on oxidising and reducing agents:

- Oxidising agents give oxygen to another substance or remove hydrogen from it.
- Reducing agents remove oxygen from another substance or give hydrogen to it.

OXIDATION AND REDUCTION IN ELECTRON TRANSFER

This is easily the most important use of the terms oxidation and reduction at A' level.

Definitions:

- *Oxidation is loss of electrons.*
- *Reduction is gain of electrons.*

It is essential that you remember these definitions. There is a very easy way to do this. As long as you remember that you are talking about electron transfer:

OIL	RIG
Oxidation is Loss	Reduction is Gain

Example

The equation shows a simple redox reaction which can obviously be described in terms of oxygen transfer.

$CuO + Mg \rightarrow Cu + Mgo$

Copper(II) oxide and magnesium oxide are both ionic. The metals obviously aren't. If you rewrite this as an ionic equation, it turns out that the oxide ions are spectator ions and you are left with:

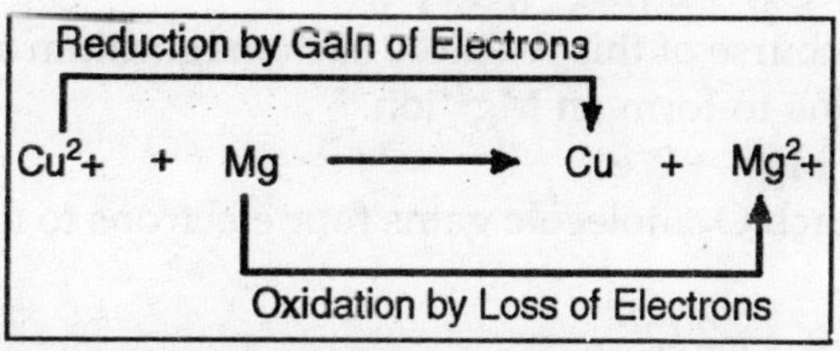

If you look at the equation above, the magnesium is reducing the copper(II) ions by giving them electrons to neutralise the charge. Magnesium is a reducing agent. Looking at it the other way round, the copper(II) ions are removing electrons from the magnesium to create the magnesium ions. The copper(II) ions are acting as an oxidising agent.

OXIDATION-REDUCTION REACTIONS

The term oxidation was originally used to describe reactions in which an element combines with oxygen.

Example: The reaction between magnesium metal and oxygen to form magnesium oxide involves the oxidation of magnesium.

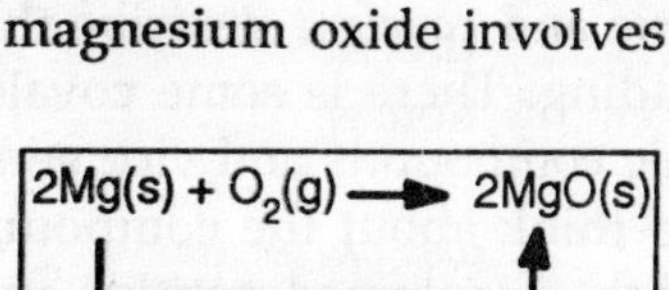

The term reduction comes from the Latin stem meaning "to lead back." Anything that that leads back to magnesium metal therefore involves reduction. The reaction between magnesium oxide and carbon at 2000C to form magnesium metal and carbon monoxide is an example of the reduction of magnesium oxide to magnesium metal.

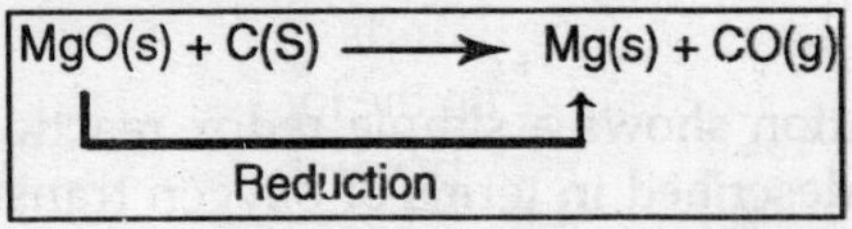

After electrons were discovered, chemists became convinced that oxidation-reduction reactions involved the transfer of electrons from one atom to another. From this perspective, the reaction between magnesium and oxygen is written as follows.

$2\ Mg + O_2 \rightarrow 2\ [Mg^{2+}][O^{2-}]$

In the course of this reaction, each magnesium atom loses two electrons to form an Mg^{2+}ion.

$Mg \rightarrow Mg^{2+} + 2^{-}e^{-}$

And, each O_2 molecule gains four electrons to form a pair of O^{2-} ions.

$O_2 + 4\ e^{-} \rightarrow 2\ O^{2-}$

Because electrons are neither created nor destroyed in a chemical reaction, oxidation and reduction are linked. It is impossible to have one without the other, as shown in the figure below.

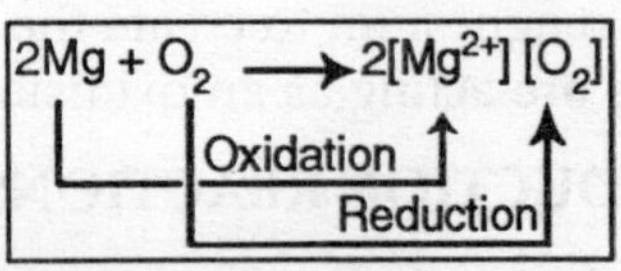

OXIDATION NUMBERS VERSUS CHARGE ON IONS

The terms *ionic* and *covalent* describe the extremes of a continuum of bonding. There is some covalent character in even the most ionic compounds and vice versa.

It is useful to think about the compounds of the main group metals as if they contained positive and negative ions. The chemistry of magnesium oxide, for example, is easy to understand if we assume that MgO contains Mg^{2+} and O^{2-} ions. But no compounds are 100% ionic. There is experimental evidence, for example, that the true charge on the magnesium and oxygen atoms in MgO is +1.5 and -1.5.

Oxidation states provide a compromise between a powerful model of oxidation-reduction reactions based on the assumption that these compounds contain ions and our knowledge that the true charge on the ions in these compounds is not as large as this model predicts. By definition, the oxidation state of an atom is the charge that atom would carry if the compound were purely ionic.

For the active metals in Groups IA and IIA, the difference between the oxidation state of the metal atom and the charge on this atom is small enough to be ignored.

The main group metals in Groups IIIA and IVA, however, form compounds that have a significant amount of covalent character. It is misleading, for example, to assume that aluminum bromide contains Al^{3+} and Br^- ions. It actually exists as Al_2Br_6 molecules.

This problem becomes even more severe when we turn to the chemistry of the transition metals. MnO, for example, is ionic enough to be considered a salt that contains Mn^{2+} and O^{2-} ions. Mn_2O_7, on the other hand, is a covalent compound that boils at room temperature. It is therefore more useful to think about this compound as if it contained manganese in a +7 oxidation state, not Mn^{7+} ions.

OXIDIZING AGENTS AND REDUCING AGENTS

Let's consider the role that each element plays in the reaction in which a particular element gains or loses electrons..

When magnesium reacts with oxygen, the magnesium atoms donate electrons to O_2molecules and thereby reduce the oxygen. Magnesium therefore acts as a reducing agent in this reaction.

$2\ Mg + O_2 \quad \rightarrow 2\ MgO$

reducingagent

The O_2 molecules, on the other hand, gain electrons from

magnesium atoms and thereby oxidize the magnesium. Oxygen is therefore an oxidizing agent.

$$2\ Mg + O_2 \rightarrow 2\ MgO$$

oxidizing agent

Oxidizing and reducing agents therefore can be defined as follows. *Oxidizing agents* gain electrons. *Reducing agents* lose electrons.

OXIDIZING AGENT/REDUCING AGENT PAIRS

Metals act as reducing agents in their chemical reactions. When copper is heated over a flame, for example, the surface slowly turns black as the copper metal reduces oxygen in the atmosphere to form copper(II) oxide.

Oxidiging Agent

$$\underset{0}{2Cu} + \underset{0}{O_2} \longrightarrow \underset{+2-2}{2CuO}$$

Reducing Agent

If we turn off the flame, and blow H_2 gas over the hot metal surface, the black CuO that formed on the surface of the metal is slowly converted back to copper metal. In the course of this reaction, CuO is reduced to copper metal. Thus, H_2 is the reducing agent in this reaction, and CuO acts as an oxidizing agent.

Oxidizing Agent

$$\underset{+2}{CuO} + \underset{0}{H_2} \longrightarrow \underset{0}{Cu} + \underset{+1}{H_2O}$$

Reducing Agent

An important feature of oxidation-reduction reactions can be recognized by examining what happens to the copper in this pair of reactions. The first reaction converts copper metal into CuO, thereby transforming a reducing agent (Cu) into an oxidizing agent (CuO).

The second reaction converts an oxidizing agent (CuO) into a reducing agent (Cu). Every reducing agent is therefore linked, or coupled, to a conjugate oxidizing agent, and vice versa.

Every time a reducing agent loses electrons, it forms an oxidizing agent that could gain electrons if the reaction were reversed.

Oxidizing Agent

$$Cu \longrightarrow Cu^{2+} + 2e^-$$

Reducing Agent

Conversely, every time an oxidizing agent gains electrons, it forms a reducing agent that could lose electrons if the reaction went in the opposite direction.

$$O_2 + 4e^- \longrightarrow 2O^{2-}$$

Oxidizing Agent Reducing Agent

The idea that oxidizing agents and reducing agents are linked, or coupled, is why they are called conjugate oxidizing agents and reducing agents. *Conjugate* comes from the Latin stem meaning "to join together." It is therefore used to describe things that are linked or coupled, such as oxidizing agents and reducing agents.

The main group metals are all reducing agents. They tend to be "strong" reducing agents. The active metals in Group IA, for example, give up electrons better than any other elements in the periodic table.

The fact that an active metal such as sodium is a strong reducing agent should tell us something about the relative strength of the Na^+ ion as an oxidizing agent. If sodium metal is relatively good at giving up electrons, Na^+ ions must be unusually bad at picking up electrons. If Na is a strong reducing agent, the Na^+ ion must be a weak oxidizing agent.

$$Na \longrightarrow Na^+ + e^-$$

Strong Reducing Agent Weak Oxidizing Agent

Conversely, if O_2 has such a high affinity for electrons that it is unusually good at accepting them from other elements, it should be able to hang onto these electrons once it picks them up. In other words, if O_2 is a strong oxidizing agent, then the O^{2-} ion must be a weak reducing agent.

$$\underset{\text{Strong Oxidizing Agent}}{O_2 + 4e^-} \longrightarrow \underset{\text{Weak Reducing Agent}}{2O^{2-}}$$

In general, the relationship between conjugate oxidizing and reducing agents can be described as follows. Every strong reducing agent (such as Na) has a weak conjugate oxidizing agent (such as the Na^+ ion). Every strong oxidizing agent (such as O_2) has a weak conjugate reducing agent (such as the O^{2-} ion).

ADVANCED OXIDATION

Advanced chemical oxidation processes make use of (chemical) oxidants to reduce COD/BOD levels, and to remove both organic and oxidisable inorganic components. The processes can completely oxidise organic materials to carbon dioxide and water, although it is often not necessary to operate the processes to this level of treatment

A wide variety of advanced oxidation processes are available:

- Chemical oxidation processes using hydrogen peroxide, ozone, combined ozone & peroxide, hypochlorite, Fenton's reagent etc.
- Ultra-violet enhanced oxidation such as UV/ozone, UV/hydrogen peroxide, UV/air
- Wet air oxidation and catalytic wet air oxidation (where air is used as the oxidant)

Advanced oxidation processes are particularly appropriate for effluents containing refractory, toxic or non-biodegradable materials.

The processes offer several advantages over biological or physical processes, including:

- Process operability
- Unattended operation
- The absence of secondary wastes
- The ability to handle fluctuating flow rates and compositions

However, advanced oxidation processes often have higher capital and operating costs compared with biological treatment. The most suitable variant for each application is chosen on the basis of the chemical properties of the effluent

OXIDATION NUMBERS

The oxidation number is the charge an atom in a substance would have if the pairs of electrons in each bond belonged to the more electronegative atom.

Now this means that in a compound made up of monatomic ions, like NaCl, in which the bonding pairs do belong to the more electronegative atom, the oxidation number equals the ionic charge. The sodium ion has an oxidation number of +1 while the chlorine ion has an oxidation number of –1.

Here are some rules that are helpful in finding the oxidation number of a specific atom:

- The oxidation number of an atom in an elementary substance is 0. This means that the oxidation number of an O atom in O_2 is 0.
- The oxidation number of a Group IA atom in any compund is +1; The oxidation number of a Group IIA atom in any compund is +2.
- The oxidation number of fluorine is –1 in all of its compounds.
- The oxidation number of chlorine, bromine, and iodine is –1 in any compound containing only two elements.
- The usual oxidation number of oxygen in a compound is –2. The major exceptions of this rule are peroxides, like H_2O_2, which have an oxidation number of –1.
- The oxidation number of hydrogen in most compounds is +1.

You might be heartily confused at this point as to what oxidation numbers are and how to find the Don't worry, you can usually find the oxidation numbers of elements in ionic compounds by looking at the individual ions in the compound.

REACTIONS

When we take an iron metal nail and dip it into a blue solution of copper (II) sulfate, the nail becomes covered with a reddish substance. This reddish susbtance is metallic copper. The equation for this is:

$Fe(s) + CuSO_4(aq) \rightarrow FeSO_4(aq) + Cu(s)$

The net ionic equation, that is, the equation in which we have cancelled out the spectator ions (in this case the spectator ion is SO_4) is:

$Fe^0(s) + Cu^{2+}(aq) \rightarrow Fe^{2+}(aq) + Cu^0(s)$

What we have done here is to show you the oxidation numbers above the elements in the equation. This shows that the iron metal loses electrons to form iron(II) ions while copper(II) ions gain electrons to create copper metal. The Cu^{2+} on the reactants side of the equation has a 2+ because SO_4 has a charge of 2–. Any reaction like this that involves a transfer of electrons is an oxidation-reduction reaction.

We can write the equation above in the form of two half-reactions. A half-reaction is one of two parts of an oxidation-reduction reaction, one of which involves a loss of electrons, and the other a gain of electrons.

The half reactions for the iron nail reaction would be:

$Fe(s) \rightarrow Fe^{+2}(aq) + 2e^-$ 2. $Cu^{+2}(aq) + 2e^- \rightarrow Cu(s)$

The first equation shows us that electrons are lost by Fe, ($2e^-$). The second equation shows electrons being gained by Cu^{2+}. You know, we can also express this reaction in terms of what happened to the oxidation number. The oxidation number of iron increases after the reaction, while the oxidation number of copper decreases.

Basically, an oxidation-reduction reaction (redox reaction) is a reaction in which electrons are transferred between species or in which atoms change oxidation numbers. Oxidation is the portion of the redox reaction in which there is a loss of electrons by a species or an increase in the oxidation number of an atom. Reduction is the part of a reaction in which there is a gain of electrons by a species or a decrease in the oxidation number of an atom. Something that is oxidized loses electrons or contains an atom that increases in oxidation number.

Something that is reduced gains electrons or has an atom that decreases in oxidation number. An oxidizing agent is a something that oxidizes something, meaning that the oxidation agent is itself reduced. A reducing agent is a something that reduces something else, and is itself oxidized. Notice the example below.

Oxidation

$$Fe(S) + Cu^{2+}(aq) \rightarrow Fe^{2+}(aq) + Cu\ (S)$$

Reuction

Fe is oxidized and Fe is the reducing agent. Cu^{+2} is reduced and Cu^{+2} is the oxidizing agent.

Chapter 4

Environment

ATMOSPHERE

The atmospheric chemistry studies the chemical composition of the natural atmosphere, the way gases, liquids, and solids in the atmosphere interact with each other and with the earth's surface.

For convenience of study, atmospheric scientists divide the atmosphere as if it consists of 4 layers. The division is mainly due to temperature variations as the altitude increases.

The 4 layers according to the variation of temperature are:

- Ionosphere
- Mesosphere
- Stratosphere
- Troposphere

In the given figure as shown below, 100 km is the thermosphere and ionosphere where the temperature increases from 200 K at 100 km to 500 K at 300 km. The temperature goes even higher as the altitude increases. activity as the altitude decrease. In the outer space, most particles consist of single atoms, H, He, and O etc. At lower altitude (200 – 100 km), diatomic molecules N_2, O_2, NO etc are present.

The ionosphere is full of electrically charged ions. The UV rays ionizes these gases.

The major reactions are:

In the ionosphere:

$O + h\nu \rightarrow O^+ + e$

$N + h\nu \rightarrow N^+ + e$

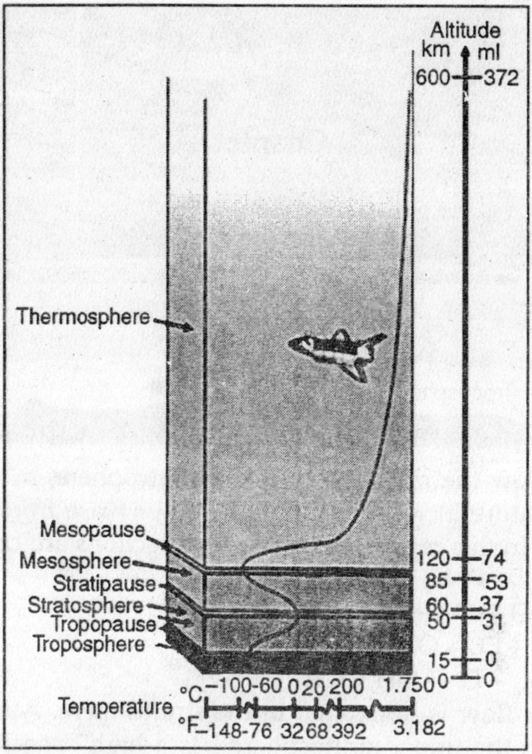

In the neutral thermosphere:

$N + O_2 \rightarrow NO + O$

$N + NO \rightarrow N_2 + O$

$O + O \rightarrow O_2$

Beyond the neutral thermosphere is the ionosphere and exosphere. These layers are of course interesting for space explorations and environmental concerns and space sciences. The atmosphere in the outer space is more like a plasma than a gas.

Below the thermosphere is the mesosphere (100 – 50 km) in which the temperature decreases as the altitude increase. In this region, OH, H, NO, HO_2, O_2, and O_3 are common, and the most prominent chemical reactions are:

$H_2O + h\,v \rightarrow OH + H$

$H_2O_2 + O \rightarrow OH + OH.$

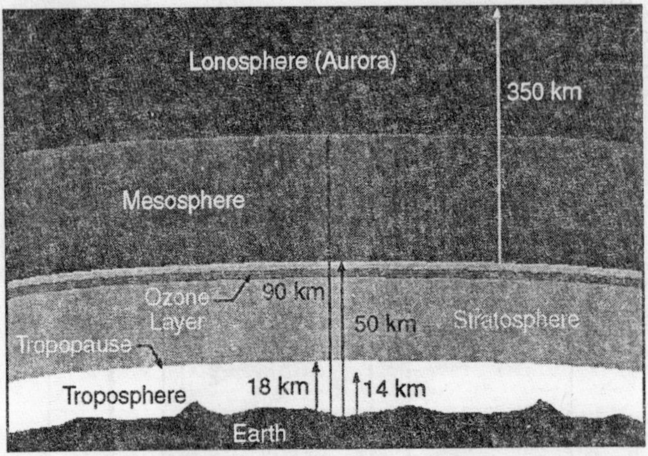

Below the mesosphere is the stratosphere, in which the temperature increases as the altitude increase from 10 km to 50 km. In this region, the following reactions are common:

$$NO_2 \rightarrow NO + O$$
$$N_2O \rightarrow N_2 + O$$
$$H_2 + O \rightarrow OH + H$$
$$CH_4 + O \rightarrow OH + CH_3$$

Air flow is horizontal in the stratoshpere. A thin ozone layer in the upper stratosphere has a high concentration of ozone. This layer is primarily responsible for absorbing the ultraviolet radiation from the sun. The ozone is generated by these reactions:

$$O_2 + h\upsilon \rightarrow O + O$$
$$O_2 + O \rightarrow O_3$$

The troposphere is where all weather takes place; it is the region of rising and falling packets of air. The air pressure at the top of the troposphere is only 10% of that at sea level (0.1 atmospheres). There is a thin buffer zone between the troposphere and the next layer called the tropopause.

The major components in the region close to the surface of the Earth are N_2 (78%), O_2 (21%), Ar (1%) with variable amounts of H_2O, CO_2, CH_4, NO_2, NO_2, CO, N_2O, and O_3. The ozone concentration in this layer is low, about 8% of the total ozone in the atmosphere is in the troposphere.

All the weather phenomena happen in the troposphere. If you want to know the current weather con dition, check the AccuWeather network for a good view of the troposphere.

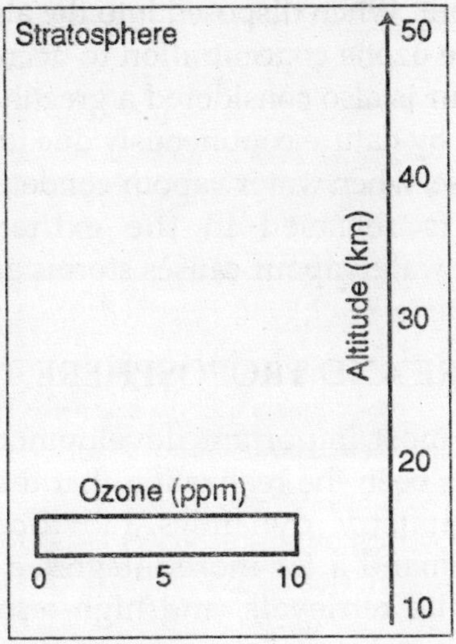

GASES IN ATMOSPHERE

From the atmospheric science viewpoint, interactions of all gasses amoung themselves and their interaction with the environmental elements are of interest. However, for identification purposes, we need to identify the gases produced by man-made process (industry).

Some of the gases due to human activities are:

- Carbon dioxide result from the excess burning of carbon-containing fuel.
- Carbon monoxide produced by automobiles. This orderless and Colourless gas is very toxic.
- Ozone produced in the exhaust of internal combustion engine, and the variation of ozone concentration in the stratosphere.
- Nitrogen oxides such as NO, NO_2, N_2O_4; due to the production of NO in the internal combustion engin.
- Methane gas produced due to treatments of large amount of waste.
- Sulfur oxides produced in mining operation and in

the combustion of sulfur containing fuel. Sulfur oxide causes the so called acid rain problem.

- Chlorofluorocarbons (CFC) are gases used as refrigurant. When disposed into the atmosphere, they cause the ozone concentration to decrease.

Water vapour is also considered a greenhouse gas, but it is also generated by nature contineously due to radiation from the Sun. Of course, when water vapour condense into a liquid, much energy is released in the exthermal process. Condensation of watervapour causes storms and many of the weather phenomena.

STRATOSPHERE AND TROPOSPHERE

One of the most important developments in the past several years has been the realization that the next phase in atmospheric chemistry, dynamics, radiation, and climate research will demand a far more integrated application of modeling, satellite retrievals, and high-resolution, in situ observations. This realization stems from two different causes:

- The first is the increasing awareness that scientific research must develop forecasts that are tested and trusted internationally in such areas as climate change, UV dosage levels resulting from ozone changes in the stratosphere, toxicity levels stemming from the urban-to-regional emission of nitrates, sulfates, organics, soot and heavy metals, etc. This demands an integration of transport and photochemical mechanisms that go well beyond the normal alliance that stems from curiosity-driven studies that operate without regard for timetables or societal demands.

The second is the increasing level of sophistication in modeling, data assimilation, and technology related to observational systems, whether they be remote or in situ. This enables the scientific community to investigate in finer granularity the cause and effect relationships in the study of atmospheric chemistry and dynamics.

In response to this new realization, the Anderson Group has pursued a course of action based upon a set of objectives of critical importance in today's world.

TROPOSPHERE AND STRATOSPHERE

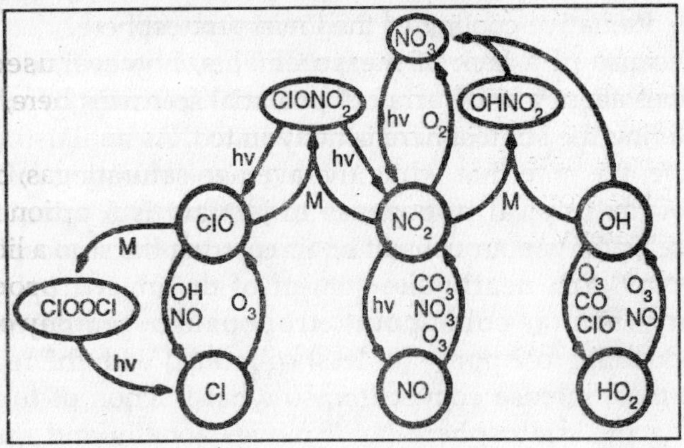

HYPOTHESIS

The boundary condition for water vapour entering the stratosphere in the tropics (irreversible passage through the 390 K è surface) is set by the vertical/horizontal temperature structure of the cold trap region over the Western Tropical Pacific through which the quasi-horizontal passage of air provides adequate residence time (coupled potentially with multiple traversals of these cold pools) to desiccate the air to a vapour pressure corresponding to the local, seasonally dependent, temperature minimum.

Understanding the processes governing stratospheric water vapour is broadly recognized as a dominating scientific need. Changes in stratospheric water vapour under future climate regimes would strongly affect both radiative balance and ozone chemistry.

In particular, increases in stratospheric water content could dramatically increase ozone losses by:

- Increasing the cold aerosol surface area that draws down the NO_x/NO_y ratio thereby increasing the fraction of reactive halogens and hydrogen species portioned into free radical form (ClO, BrO, IO, OH and HO_2)
- Promoting formation of the polar stratospheric clouds on which catalytic ozone loss chemistry occurs

- Increasing the source of HO_x radicals
- Radiative cooling of the lower stratosphere

Because of a lack of measurements, however, debate continues about a wide array of potential scenarios by which air entering the stratosphere is dehydrated. As an illustrative example, we note that while the average saturation mixing ratio of the tropical tropopause in January is 3 ppmv, the average water vapour content of air entering the stratosphere is 4 ppmv. Yet, nearly one quarter of the inner tropics in January has a cold point tropopause temperature corresponding to 2 ppmv (or less) co-located with the region of the most intense convection; so what fraction of the air entering the stratosphere (as it moves zonally and quasi-horizontally with convection superimposed on this background motion) has passed through this cold region?

OZONE AND UV DOSAGE

Societal Objectives

The mission statement identifies as a first order societal issue: "To understand and protect our home planet." A central challenge facing the United States and other countries in the twenty-first century will be to enhance human well-being in a world where growing populations and the drive to improve living standards place potentially huge demands on natural resources and the environment.

Whether we succeed or fail in meeting this challenge will be determined, in part, by how we respond to immediate demands to address human health and economic growth in the context of the wide range of crucial, environmentally related decisions made every day by insurance companies, water resource managers, agribusiness, households, city planners, public health officials, and countless others. Rising to this challenge will entail using natural resources as efficiently as possible, devising practical solutions that meet our immediate needs and also provide for long-run economic growth, while maintaining the environmental systems on which life depends.

As we have seen, the Earth's system can be fragile.

Stratospheric ozone loss has had profound implications for human health. It has lead to disfiguring and deadly skin cancers that occur as a result of increased ultra-violet radiation reaching the earth's surface.

Losses over northern mid-latitudes will affect the population Centres of Europe and North America. Isolating the cause is complicated by the fact that the photochemical lifetime of ozone in this region is comparable to the dynamical residence time and the dynamical residence time is highly variable depending on season, altitude, longitude, and latitude. Careful measurements and analyses are required to identify and quantify the role of both chemistry and dynamics in determining mid-latitude ozone concentrations.

To guide wise public policy decisions that continue to improve human and economic conditions and to clarify public debate, it is necessary to restructure the science and engineering framework addressing the biological, chemical, and physical integrity of our surroundings. Private-sector and governmental decisions will be made regarding air, water, and living systems that will fundamentally affect our nation's health and its economic and environmental vitality.

AIRBORNE MISSIONS

Airborne Platform

The Anderson Group instruments constitute an important component of the observational array required to address both the AURA calibration/validation requirements and the hypotheses that constitute the scientific objectives of the Programme. Consideration of AURA colLabourative science and the scientific objectives of TC^4 serves to identify required characteristics of the platform used in the airborne field operation that constitutes the core of this effort:

- Ability to carry the proper complement of instruments
- Ability to contiguously cover the altitudes required from the boundary layer to the lower stratosphere
- Ability to operate in convectively active regions and to engage in sophisticated storm avoidance

- Ability to minimize operational complexity for foreign deployments
- Ability to accept new instruments and to accept multiple instruments for intercomparison
- Ability to combine *in situ* and remote observations to increase the colLabourative overlap between aircraft and satellites.

LASER DEVELOPMENT

A problem of fundamental importance to the NASA vision and mission to improve life here and to understand and protect our home planet, involves a new strategic plan for establishing the quantitative link between rapidly increasing urban/ regional emission of aerosol precursors (nitrate, sulfate, heavy metal, organic and elemental carbon) and changes in the optical properties of the atmosphere that constitute a powerful climate forcing. Dissecting the mechanisms underlying these changes requires measurements that span temporal scales from seconds to years and spatial scales from tens of meters to thousands of Kilometres, to provide an effective union with A-Train satellites. This research will advance an array of innovative new developments in cavity enhanced spectroscopic techniques employing new miniaturized laser systems coupled with high finesse cavities that will provide a new generation of observations from NASA UAVs.

Central to these efforts will be the advancement of:

- Difference frequency generation (DFG), both singly resonant and intracavity,
- Pulsed fibre lasers using emerging master oscillator power amplifier (MOPA) configurations, micro-etalons, and double-clad fibers, and
- Tailored quantum cascade lasers using newly developed beam epitaxy approaches.

In conjunction with these laser developments, we will advance developments in flight electronics with emphasis on scalable distributed processing and data acquisition necessary for the inclusion of a scientifically potent instrument array within the weight and volume restrictions of UAVs. A key component of this research endeavor is the objective of unifying the sub-orbital and satellite components of the Programme.

COMBUSTION RELATED AIR POLLUTION

AIR POLLUTION

Air pollution, both indoor and outdoor, is a significant cause of health problems worldwide. Urban and rural outdoor environments contain infections, allergens, irritants and chemical toxins that can reduce the quality of life and cause disease. Inhaled air pollution is directed at the the nose, throat and lungs. The exposed airway allows hazardous pollutants to enter the body and all tissues are ultimately exposed.

The tolerance for environmental destruction is ancient and human history is littered with civilizations that failed because humans indiscriminately exploited natural resources and spoiled their own nest. Humans adapt easily to deteriorating conditions and will persist in following daily routines even when air pollution is severe, traffic is congested, water and food supplies are at risk, and social order is unstable.

Fine-particulate air pollution is known to contribute to cardiovascular and lung disease, increasing the risk of heart attack and a heart-related death. Researchers at Brigham Young University and Harvard School of Public Health compared changes in air pollution from 1980 to 2000 with residents' life expectancies, They concluded that a reduction in air pollutants in 51 U.S. cities between 1980 and 2000 added an average of five months to life expectancy.

Residents in cities that made the most significant improvements in air quality, such as Pittsburgh, PA, lived almost 10 months longer. For every microgram per cubic meter decrease in fine-particulate air pollution, life expectancies rose by more than seven months.

Airborne chemicals contaminate food and water. They may be ingested and are also collected in the nose and throat and swallowed, often in mucus that attempts to protect exposed surfaces. Airborne chemicals entering the digestive system include well-known toxins such as pesticides, organophosphate, PCBs, dioxin, arsenic, cadmium, lead, and mercury. In addition occupational exposures to airborne pathogens can be intense and can cause cancer.

While ambient air pollution is a major concern, indoor air can be more polluted than outdoor air. Building materials and furnishings are a source of volatile chemicals. A decrease in indoor air quality is the result of reduced ventilation and efficient construction practices, sealing homes, stores and office buildings from the outdoor environment

Smoking tobacco remains a personal method of producing air pollution that remains popular worldwide, despite overwhelming evidence that tobacco smoke produces a long list of disabling and fatal diseases. It is estimated that 30% of all fatal cancers could be prevented if tobacco smoking were eliminated from the list of air pollutants.

The really sad part of our current predicament is that all the right ideas for creating a healthy environment have been around for decades and have been clearly articulated in many forms by a host of intelligent people. The right ideas involve unselfish and compassionate Behaviour. The right ideas involve long-term planning, conservation and deep commitment to preserving the natural world. Without a healthy natural environment, there will be few or no healthy humans.

Our big environmental problems are built from many small, personal decisions - little mistakes that add up over time. If there is a solution, it will emerge from the collective value of millions of better decisions made by individuals all over the globe. The environmental action plan is to think globally and act locally - it does make sense.

AIR POLLUTANTS

In cities across the globe, the personal automobile is the single greatest polluter, as emissions from a billion vehicles on the road add up to a planet-wide problem. Driving a private car is a typical citizen's most air polluting activity. The negative effects of automotive emissions are maximum when you sit in traffic surrounded by cars, their engines idling. Everyone in a traffic jam is getting poisoned.

The Combustion Process Gasoline and diesel fuels are mixtures of hydrocarbons (made of hydrogen, oxygen and carbon atoms.) Hydrocarbons are burned by combining with oxygen. Nitrogen and sulphur atoms are also present and combine with oxygen when burned to produce gases. Automotive engines emit several types of pollutants.

Typical Engine Combustion

Fuel + Air ⇒ Hydrocarbons + Nitrogen Oxides + Carbon Dioxide + Carbon Monoxide + water

Hydrocarbon emissions are fragments of fuel molecules, only partially burned. See Toxicity of Benzene and other Hydrocarbons in exhaust.

Hydrocarbons react in the presence of nitrogen oxides and sunlight to form ground-level ozone, a major component of smog. Ozone irritates the eyes, nose, throat and damages the lungs. A number of exhaust hydrocarbons are also toxic, some with the potential to cause cancer.

Nitrogen Oxides

Under high pressure and temperature conditions in an engine, nitrogen and oxygen atoms react to form nitrogen oxides. Nitrogen oxides, like hydrocarbons, are precursors to the formation of ozone and contribute to acid rain. Catalytic converters in car exhaust systems break down heavier nitrogen gases, forming nitrous oxide (NO_2) - 300 times more potent than carbon dioxide as a greenhouse gas. Nitrous oxide makes up about 7.2 per cent of the gases that cause global warming. Vehicles with catalytic converters produced nearly half of that nitrous oxide. Nitrous oxide also comes from nitrogen-based fertilizers and manure from farm animals.

Carbon Monoxide

Carbon monoxide (CO) is a Colourless, odorless, poisonous gas, a product of incomplete burning of hydrocarbon-based fuels.

Carbon monoxide consists of a single carbon atom and a single oxygen atom linked together (CO), the product Carbon monoxide of incomplete combustion of fuel. Most CO is produced when air-to-fuel ratios are too low in the engine during vehicle starting, when cars are not tuned properly, and at higher altitudes, where thin air reduces the amount of oxygen available for combustion.

Two-thirds of the carbon monoxide emissions come from transportation sources, with the largest contribution coming from cars. In urban areas, the passenger vehicle contribution to carbon moncxide pollution can exceed 90%.

Carbon Dioxide

Environmental Protection Agency (EPA) originally viewed carbon dioxide as a product of "perfect" combustion, but now views CO_2 as a pollution concern. Carbon dioxide is a greenhouse gas that traps the earth's heat and contributes to Global Warming.

Evaporative Emissions

Hydrocarbon pollutants also escape into the air through fuel evaporation - evaporation causes significant hydrocarbon pollution from cars on hot days when ozone levels are highest. Evaporative emissions occur several ways:

Diurnal

Gasoline evaporation increases as the temperature rises during the day, heating the fuel tank and venting gasoline vapors.

Running Loses

The hot engine and exhaust system can vaporize gasoline when the car is running.

Sitting Evaporation

The engine remains hot for a period of time after the car is turned off, and gasoline evaporation continues when the car is parked.

Adding Fuel

Gasoline vapors are always present in fuel tanks. These vapors are forced out when the tank is filled with liquid fuel.

Benzene is the main toxin in the hydrocarbon fraction of exhaust. Benzene and other less known hydrocarbons are produced in petroleum refining, and are widely used as solvents and as materials in the production of various industrial products and pesticides. Benzene also is found in gasoline and in cigarette smoke. Other environmental sources of benzene include gasoline (filling) stations, underground storage tanks that leak, wastewater from industries that use benzene, chemical spills, and groundwater next to landfills containing benzene. Exposure to benzene is related to the development of leukemia and lymphoma. Benzene has a suppressive effect on bone marrow and it impairs blood cell maturation and amplification.

Polycyclic Aromatic Hydrocarbon (PAH)

PAHs are a group of chemicals that are formed during the incomplete burning of coal, oil and gas, garbage, or other organic substances. PAHs can be man-made or occur naturally. A few of the PAHs are used in medicines and to make dyes, plastics, and pesticides. They are found throughout the environment in the air, water and soil. There are more than 100 different PAH compounds. Although the health effects of the individual PAHs vary, the following 15 PAHs are considered as a group with similar toxicity: acenaphthene, acenaphthylene, anthracene, benzanthracene, benzopyrene, benzofluoranthene, benzoperylene, benzofluoranthene, chrysene dibenzanthracene, fluoranthene, fluorene, indenopyrene, phenanthrene, pyrene.

Long term solutions require reduced combustion of all kinds. While vehicles with new energy sources such ethanol, biofuels, propane and natural gas can contribute to reduced air pollution, their benefit is limited if vehicle use continues at current intensities. If you pay more money to buy a hybrid car, but drive it more, you have contributed little to solving air pollution problems. If you buy a gas guzzling clunker and use only one gallon of gas to 20 miles each week, you have contributed more to the solution.

The problem with all alternative fuels is that the manufacture of fuels requires energy, distribution with a manufacturing infrastructure that consume energy, often derived from burning fossil fuels. No alternative fuel is ideal. While ethanol has been championed as an alternative to petroleum fuels, it mainly helps to reduce dependency on oil producing countries.

When ethanol is made from corn, more than 75% of its energy value must be spent on its production. Burning ethanol still produces carbon dioxide. Climate change with extreme weather events threatens corn production in the US. The new competition between hastily constructed ethanol plants and food production became an international issue in 2008. Other approaches to new fuels made from non-food plant sources are part of the solution.

Hydrogen

Ultimately cars might burn hydrogen in fuel cells, but despite working prototypes, a hydrogen economy is a distant fantasy. There are many problems to be solved before hydrogen can replace fossil fuels as a portable energy source. The biggest problem is that producing hydrogen requires a large amount of energy.

In Canada, there are opportunities to dam rivers and produce electricity with falling water, a non polluting, renewable energy resource. A more problematic energy source would be be nuclear reactors that"burn"uranium or plutonium. Even if new non-polluting energy sources are developed, hydrogen storage and distribution requires a new, very expensive infrastructure that could replace gasoline and diesel fuels. With once rich countries such as the USA on the verge of bankruptcy and facing the extensive repairs of already aging, derelict infrastructures, adding a new, unprecedented development costs seems unlikely. Unless, of course the priorities in the US shift dramatically. The US could adopt a sane, smart strategy, reduce its military budget by 60% and invest the money and skills in rebuilding the country's infrastructure with new sustainable energy sources.

GLOBAL ENVIRONMENTAL PROBLEMS

INTRODUCTION

Ethics is understood to be the domain of inquiry that explores what is right or wrong, obligatory or non-obligatory, or when responsibility attaches to human Behaviour. Are there features of global environmental problems that call for classifying them as essentially ethical problems with even greater force than some local or regional environmental problems? If so, what are these features?

Why is this important? If some global environmental problems are essentially ethical problems, then some of the excuses that nations often use to justify not reducing their contributions to the global environmental problems are ethically problematic. This is so because ethical obligations entail duties and responsibilities to others with the result that national policies may not be justified on national interests alone. That is, if there are ethical obligations to others to cease causing harm, then policy options must be responsive to these obligations.

The features of global environmental problems that strongly call for classification as ethical problems. Some global environmental problems are caused by people in one part of the world but are most harshly experienced by others who are separated from those causing the problem by great space and time. In addition, those most vulnerable to global environmental problems are often least responsible for causing them. Climate change is a strong example of this feature of global environmental problems. Those most vulnerable to climate include many of the poorest people in the world who are among those least responsible for emitting the greenhouse gases that are causing the problem.

The same separation between victim and cause is also true of other global environmental problems, including upper atmospheric ozone depletion, some ocean degradation problems including destruction of global fisheries, some biodiversity and forest problems, and exposure to toxic substances that are deposited around the world by long-range air transport.

In the case of global fisheries depletion, for instance, fishing trawlers from some countries often go great distances to harvest fish for domestic consumption, yet the harshest consequences of depleted global fisheries are often experienced by subsistence fisherman in poor countries far away from the home port of the fishing boats.

This separation of space and time and between cause and impact of course can also happen in more local environmental problems such as water pollution, but the distances between those causing the problem and those who suffer the consequences are most problematic at the global scale because of the magnitude of the distances at this scale.. With the rise in truly global environmental problems in the last few decades, it is now abundantly clear that Behaviour of people in one part of the world can gravely affect the human health, economic viability and overall welfare of others far away, but their suffering is almost invisible to those who cause the problem.

Several implications for ethical reasoning follow directly from this separation of distance and time between cause and effect. First, our intuitive sense of ethical responsibility for harms that we cause others is harder to trigger when those who will be harmed by our action are people we will not ever know and are from a different culture. Yet ethics and morality require us to refrain from harming all others who may suffer because of our Behaviour even if they are separated from us in time and space.

Secondly, because the actual harm that we are causing to others may be in a particular part of the world that is uniquely vulnerable to global environmental change but is unknown to us, we not only don't know the people we are harming but have no idea where the locus of the worst harm is located. For instance, particularly vulnerable villages to climate change impacts may turn on the geographical features where the village is located such as dependence on glacier water flows in the summer coupled with vulnerability to monsoon irregularities. Yet few of us know where these places are. And so in the case of global environmental problems we may be causing great harm to others who we not only do not know personally, but about whom we neither have an idea of their location nor why they are particularly vulnerable to actions we are taking. For this reason, global scale environmental problems create a huge challenge to trigger our sense of ethical responsibility, yet ethics requires us to not harm others through our actions none the less.

A related reason why the global environmental problems demand classification as ethical problems stems from the fact that some actions that cause some environmental damage are often justified on the basis of cost-benefit analysis. Yet this justification is ethically problematic when those who benefit from an action that causes harm have no relation with those who are harmed. For instance, government may tolerate inevitable pollution from a land fill by identifying the benefits that citizens receive by having a place to take their garbage. But in the case of global environmental problems, those who are harmed by actions do not enjoy the benefits that come from taking the action. A harms-benefits justification for environmental degradation, if ethically justifiable at all, only works where those who will suffer the harms and those who benefit from the action that causes the harm are in the same community of interests. In the case of global environmental problems, those who are harmed are rarely made better by the benefits of the action that cause the harm.

The second reason why global environmental problems call for classification as ethical problems stems from the fact that their consequences are often catastrophic to those who are most harmed by them. Climate change, for instance, directly threatens human life and health and resources to sustain life as well as species of plants and animals and ecosystems around the world. The harms include deaths from disease, droughts, floods, heat and intense storms and damage to homes and villages from rising oceans and intense storms, adverse impacts on agriculture, social disputes caused by diminishing natural resources, sickness from a variety of diseases, the inability to rely upon traditional sources of food, the destruction of water supplies, and the inability to live where one has lived to sustain life. In addition, the very existence of some small island nations is threatened by climate change caused seal level rise. Clearly these impacts are catastrophic for some.

Loss of global fisheries can be devastating to substance fishermen, global desertification and drought are likely to cause mass starvation, loss of atmospheric ozone can cause skin cancer in millions, and long-range transport of persistent organic chemicals can cause deadly diseases and birth defects. Because ethics requires people to be particularly careful to not to harm others when the harm is great, global environmental problems need to be seen as ethical issues.

Governments are expected to protect their citizens from life-threatening dangers. At the local, regional or national scale, citizens can petition their governments to protect them from environmental harms. But at the global level, no government exists whose jurisdiction matches the scale of global environmental problems. And so, although existing national, regional and local governments have jurisdiction over activities within their boundaries coupled to responsibilities to their citizens, they have no responsibility to those outside their boarders in the absence of binding international law. For this reason, ethical appeals to get governments to require that their citizens not harm non-citizens in other nations are necessary to prompt action.

In fact, according to a leading foreign policy theory called "realism," national governments should limit the scope of their international engagements to the pursuit of national interest alone. According to some "realist" proponents, governments should refrain from acting at the international level for ethical reasons that do not coincide with national interests. Yet global environmental problems require people in one part of the world to consider the interests of others outside the jurisdiction of their national governments. Because there is no duty of national governments to protect foreigners from global environmental problems unless they bind themselves voluntarily in treaties, an appeal to ethical and moral responsibility is particularly important to motivate national actions in regard to global scale environmental problems. Moreover, if global environmental problems actually create ethical obligations, they have ethical duties to mitigate global environmental problems

CFCS IN ATMOSPHERE

There has been a movement since the late 1970s to ban CFCs because of their destructive effect on the ozone layer. This damage was discovered by Sherry Rowland and Mario Molina, who first published a paper suggesting the connection in 1974. It turns out that one of CFCs' most attractive features-their unreactivity-has been instrumental in making them one of the most significant pollutants. CFCs' lack of reactivity gives them a lifespan which can exceed 100 years in some cases. This gives them time to diffuse into the upper stratosphere. Here, the sun's ultraviolet radiation is strong enough to break off the chlorine atom, which on its own is a highly reactive free radical. This catalyses the break up of ozone into oxygen:

$$Cl + O3 \rightarrow ClO + O_2$$

$$ClO + O \rightarrow Cl + O_2$$

CFCs are a problem because the chlorine is regenerated at the end of these reactions, making it able to repeat this reaction millions of times. This disrupts the ionic structure of oxygen atoms, causing their lowest ionisation energy to drop. This reaction is believed to be the cause of the ozone hole observed over the poles and upper latitudes of the Earth. Decreases in stratospheric ozone have been linked to increases in skin cancer.

In 1975, Oregon enacted the world's first ban of CFCs. By 1985, scientists observed a dramatic seasonal depletion of the ozone layer over Antarctica, which led the United Nations to take action in 1987. International attention to CFCs resulted in a meeting of world diplomats in Montreal in 1987. They forged a treaty, the Montreal Protocol, which called for drastic reductions in the production of CFCs. On March 2, 1989, 12

European Community nations agreed to ban the production of all CFCs by the end of the century. In 1990, diplomats met in London and voted to significantly strengthen the Montreal Protocol by calling for a complete elimination of CFCs by the year 2000. By the year 2010 CFCs should be completely eliminated from developing countries as well

FREON

Freon is a trade name for a group of chlorofluorocarbons used primarily as a refrigerant. The word Freon is a registered trademark belonging to Dupont. Prior to their implication as a major contributor to ozone depletion (and a minor contributor to the greenhouse effect), freons were also used as aerosol propellants and to foam up polymer materials.

Freon was developed as a safer alternative to toxic gases such as, ammonia (NH_3), methyl chloride (CH_3Cl), and sulfur dioxide (SO_2), that were used as refrigerants prior to the discovery of freon. Despite the controversy, R12 works well and is very safe. Its inventor, Thomas Midgley, actually breathed in a lungful and exhaled over a candle to demonstrate its nontoxicity and nonflammability.

Before CFCs were banned by developed nations, a 12 ounce (340 g) can of freon (R12) cost about \$1; following the ban the cost of freon rose dramatically. People now go through considerable trouble to recycle freon from existing sources. As of 2002, freon cost about \$85 per lb (\$187 per kg). Freon is most frequently sold in 12 ounce (340 g) cans and 30 pound (13.6 kg) canisters.

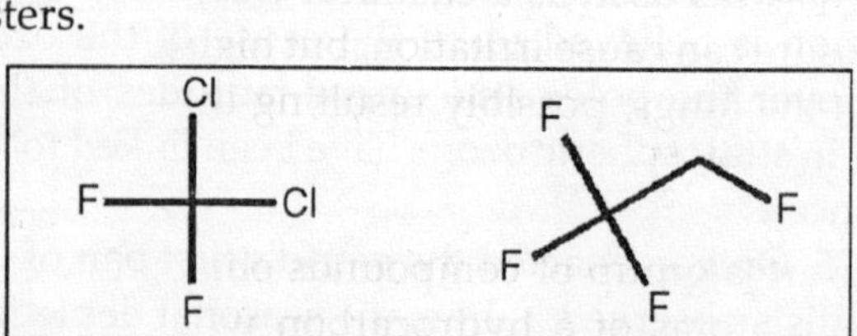

Dichlordifuoromethane "Freon 12" 1,1,1,2-Tetrafluoroethane "Suva 134a"

Freon 12 → (also known as R–12) was a popular refrigerant before it was banned. The chemical formula is $CCl_2F_2$and its proper chemical name is dichlorodifluoromethane. When people refer to freon without specifying a number, it is likely that they mean freon-12.

Suva 134a → (also known as R–134 or R–134a) is the most common replacement for R–12. Unlike R–12, which is a chlorofluorocarbon, R–134 is a hydro fluorocarbon. The chemical formula is CH_2FCF_3 and the proper chemical name 1,1,1,2-tetrafluoroethane.

	Freon 41	**fluoromethane**
R-11	CFC-11	trichlorofluoromethane
R-12	CFC-12	dichlorodifluoromethane
R-13	CFC-13	chlorotrifluoromethane
R-22	HCFC-22	chlorodifluoromethane
R-23	HFC-23	trifluoromethane
R-113	CFC-113	trichlorotrifluoroethane
R-114	CFC-114	1,2-dichloro-1,1,2,2-tetrafluoroethane
R-115	CFC-115	1-chloro-1,1,2,2,2-pentafluoroethane
R-116	CFC-116	hexachloroethane
R-134a	HFC-134a	1,1,1,2-tetrafluoroethane
R-227ea	HFC-227ea	1,1,1,2,3,3,3-heptafluoropropane

Freon in copper tubing open to the environment can turn into phosgene gas after coming in contact with extreme heat, such as, while brazing or in a fire situation. Phosgene is a substance that was used as a chemical weapon in World War I. Low exposure can cause irritation, but high levels cause fluid to collect in the lungs, possibly resulting in death.

HALON

Halons are a group of compounds obtained by replacing the hydrogen atoms of a hydrocarbon with halogen atoms, such as bromine or fluorine. Halon 1211 is bromochlorodifluoromethane (CF_2BrCl) and Halon 1301 is bromotrifluoromethane (CF_3Br). Halons are very stable and are widely used in fire extinguishers where water and other alternatives would be ineffective and dangerous (e.g. when dealing with oil fires) or cause unacceptable collateral damage (e.g. with electronic equipment.)

Introduction

At high temperatures, halons decompose to release halogen atoms that combine readily with active hydrogen atoms, quenching the flame propagation reaction even when adequate fuel, oxygen and heat remains. The chemical reaction in a flame proceeds as a free radical chain reaction; by sequestering the radicals which propagate the reaction, halons are able to "poison" the fire at much lower concentrations than are required by fire suppressants using the more traditional methods of cooling, oxygen deprivation, or fuel dilution.

For example, Halon 1301 flood systems are typically used at concentrations no higher than 6% v/v in air, and can suppress many fires at 2.9% v/v. By contrast, carbon dioxide fire suppression flood systems are operated from 34% (surface-only combustion of liquid fuels) up to 75% (dust traps). Carbon dioxide can cause severe distress at concentrations of 3 to 6%, and has caused death by respiratory paralysis in a few minutes at 10% concentration.

Halon 1301 causes only slight giddiness at its effective concentration of 5%, and even at 15% persons remain conscious but impaired, and suffer no long term effects. (Experimental animals have also been exposed to 2% concentrations of Halon 1301 for 30 hours per week for 4 months, with no discernible health effects at all.) Halon 1211 also has low toxicity, although it is more toxic than Halon 1301, and thus considered unsuitable for flooding systems.

Fig. Halon Canisters Used in a Fire-suppression System

However, halon fire suppression is not completely non-toxic; very high temperature flame, or contact with red-hot metal, can cause decomposition to toxic byproducts. The presence of such byproducts is readily detected because they include hydrobromic acid and hydrofluoric acid, which are intensely irritating.

Halons are very effective on Class A (organic solids), B (flammable liquids and gases) and C (electrical) fires, but they are totally unsuitable for Class D (metal) fires, as they will not only produce toxic gas and fail to halt the fire, but in some cases pose a risk of explosion. Halons can be used on Class K (kitchen oils and greases) fires, but offer no advantages over specialised foams.

Nomenclature

Halons are referred to by a system of code numbers similar to (but simpler than) the system used for freons. The first digit specifies the number of carbon atoms in the molecule, the second is the number of fluorine atoms, the third is the chlorine atoms, and the fourth is the number of bromine atoms. If the number includes a fifth digit, the fifth number indicates the number of iodine atoms (though iodine in Halon is rare). (Note that the various atomic species are listed in order of increasing atomic number.) Any bonds not taken up by halogen atoms are then allocated to hydrogen atoms.

For example, consider Halon 1211:

```
C F Cl Br
1 2 1  1
```

Halon 1211 has one carbon atom, two fluorine atoms, one chlorine atom and one bromine atom. A single carbon only has four bonds, all of which are taken by the halogen atoms, so there is no hydrogen. Thus its formula is CF_2BrCl, and its lUPAC name is therefore bromochlorodi fluoromethane.

Application

Halon 1211 is typically used in hand-held extinguishers, in which a stream of liquid halon is directed at a smaller fire by a user. The stream evaporates under reduced pressure, producing strong local cooling, as well as a high concentration of halon in the immediate vicinity of the fire. In this mode, extinguishment is achieved by cooling and oxygen deprivation at the core of the fire, as well as radical quenching over a larger area. After fire suppression, the halon moves away with the surrounding air, leaving no residue.

Halon 1301 is more usually employed in flooding systems. In these systems, banks of halon cylinders are kept pressurised to about 4 MPa with compressed nitrogen, and ducts lead to the protected enclosure. On triggering, the entire measured contents of one or more cylinders are dumped into the enclosure in a few seconds, through nozzles designed to ensure uniform mixing throughout the room. The quantity dumped is pre-calculated to achieve the desired concentration, typically 3-5% v/v. This level is maintained for some time, typically twenty minutes, to ensure all items have cooled so reignition is unlikely to occur, then the air in the enclosure is purged. During this time the enclosure may be entered by persons wearing SCBA. (There exists a common myth that this is because halon is highly toxic; in fact it is because it can cause giddiness and mildly impaired perception, and also due to the risk of combustion byproducts.)

Flooding systems may be manually operated or automatically triggered by a VESDA. In the latter case, a warning siren and strobe lamp will first be activated for a few seconds to warn personnel to evacuate the area. The rapid dumping of halon and consequent rapid cooling fills the air with fog, and is accompanied by a loud, disorienting noise.

Environmental Issues

There is concern that halons are being broken down in the atmosphere to bromine, which reacts with ozone, leading to depletion of the ozone layer (this is similar to the case of chlorofluorocarbons such as freon). These issues are complicated: the kinds of fires that require halon extinguishers to be put out will typically cause more damage to the ozone layer than the halon itself, not to mention human and property damage. However, fire extinguisher systems must be tested regularly, and these tests may lead to damage. As a result, some regulatory measures have been taken, and halons are being phased out in most of the world.

History

Carbon tetrachloride was used in fire extinguishers and glass (anti)-"fire grenades"from the latenineteenth century until around the end of World War II. Experimentation with halogenated hydrocarbons for fire suppression on military aircraft began at least as early as the 1920s. During World War II, various early halons were in standard use in aircraft by some combatants, but these early halons suffered from excessive toxicity. Nevertheless after the war they slowly became more common in civil aviation as well.

In the 1960s, fluorinated halons became available, and were quickly recognised as one of the most effective fire fighting materials discovered. Much early research with Halon 1301 was conducted under the auspices of the US Armed Forces, while Halon 1211 was initially mainly developed in the UK. By the late 1960s, they were standard in many applications where water and dry powder extinguishers posed a threat of damage to the protected property, including computer rooms, telecommunications switches, Labouratories, museums and art collections. Beginning with warships in the 1970s, halons also progressively came to be associated with rapid knockdown of severe fires in confined spaces with minimal risk to personnel.

By the early 1980s, halons were in common use on aircraft, ships and large vehicles, as well as in computer facilities and galleries. However, concern was beginning to be felt about the possible impact of chlorinated and brominated hydrocarbons on the ozone layer. The Vienna Convention on Ozone Layer Protection did not cover halons, as it was felt that emergency discharge of systems was too small in volume to produce a significant impact, and too important to human safety for restriction. However, by the time of the Montreal Protocol it was realised that discharges during system tests and maintenance accounted for substantially larger volumes than emergency discharges, and so halons were brought into the treaty, but with many exceptions.

Halons have been generally phased out and the possession of halon equipment is prohibited in some countries like the Netherlands and Belgium from January 1, 2004, based on the Montreal Protocol and guidelines of the European Union. Production of new stocks has ceased in most (probably all) countries as of 1994. However many countries still require aircraft to be fitted with halon fire suppression systems, as no safe and completely satisfactory alternative has been discovered for this application. There are also a few other highly specialised users.

OZONE DEPLETION

The ozone layer protects the Earth from the ultraviolet rays sent down by the sun. If the ozone layer is depleted by human action, the effects on the planet could be catastrophic.

Ozone is present in the stratosphere. The stratosphere reaches 30 miles above the Earth, and at the very top it contains ozone. The suns rays are absorbed by the ozone in the stratosphere and thus do not reach the Earth.

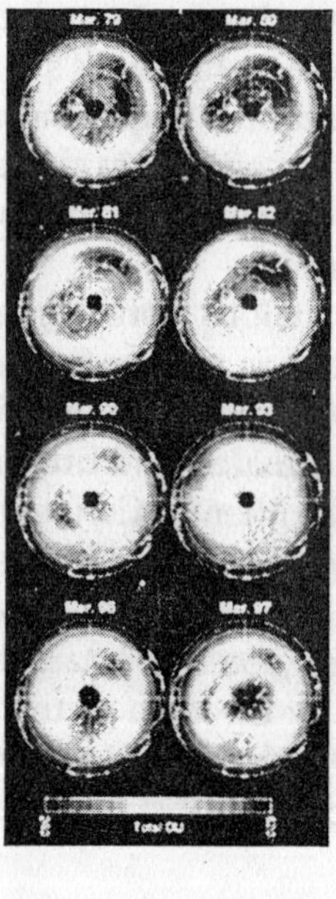

Ozone is a bluish gas that is formed by three atoms of oxygen. The form of oxygen that humans breathe in consists of two oxygen atoms, O_2. When found on the surface of the planet, ozone is considered a dangerous pollutant and is one substance responsible for producing the greenhouse effect.

The highest regions of the stratosphere contain about 90% of all ozone. In recent years, the ozone layer has been the subject of much discussion. And rightly so, because the ozone layer protects both plant and animal life on the planet.

The fact that the ozone layer was being depleted was discovered in the mid-1980s. The main cause of this is the release of CFCs, chlorofluorocarbons.

Antarctica was an early victim of ozone destruction. A massive hole in the ozone layer right above Antarctica now threatens not only that continent, but many others that could be the victims of Antarctica's melting icecaps. In the future, the ozone problem will have to be solved so that the protective layer can be conserved.

CAUSES

Only a few factors combine to create the problem of ozone layer depletion. The production and emission of CFCs, chlorofluorocarbons, is by far the leading cause.

Many countries have called for the end of CFC production because only a few produce the chemical. However, those industries that do use CFCs do not want to discontinue usage of this highly valuable industrial chemical.

CFCs are used in industry in a variety of ways and have been amazingly useful in many products. Discovered in the 1930s by American chemist Thomas Midgley, CFCs came to be used in refrigerators, home insulation, plastic foam, and throwaway food containers. Only later did people realise the disaster CFCs caused in the stratosphere. There, the chlorine atom is removed from the CFC and attracts one of the three oxygen atoms in the ozone molecule. The process continues, and a single chlorine atom can destroy over 100,000 molecules of ozone. In 1974, Sherwood Rowland and Mario Molina followed the path of CFCs. Their research proved that CFCs were entering the atmosphere, and they concluded that 99% of all CFC molecules would end up in the stratosphere.

When the ozone layer hole was discovered over Antarctica, was the proof truly conclusive. At that point, it was hard to question the destructive capabilities of CFCs.

Even if CFCs were banned, problems would remain. There would still be no way to remove the CFCs that are now present in the environment. Clearly though, something must be done to limit this international problem in the future

EFFECTS

Even minor problems of ozone depletion can have major effects. Every time even a small amount of the ozone layer is lost, more ultraviolet light from the sun can reach the Earth.

Every time 1% of the ozone layer is depleted, 2% more UV-B is able to reach the surface of the planet. UV-B increase is one of the most harmful consequences of ozone depletion because it can cause skin cancer. The increased cancer levels caused by exposure to this ultraviolet light could be enormous. The EPA estimates that 60 million Americans born by the year 2075 will get skin cancer because of ozone depletion. About one million of these people will die.

In addition to cancer, some research shows that a decreased ozone layer will increase rates of malaria and other infectious diseases. According to the EPA, 17 million more cases of cataracts can also be expected.

The environment will also be negatively affected by ozone depletion. The life cycles of plants will change, disrupting the food chain. Effects on animals will also be severe, and are very difficult to foresee. Oceans will be hit hard as well. The most basic microscopic organisms such as plankton may not be able to survive. If that happened, it would mean that all of the other animals that are above plankton in the food chain would also die out. Other ecosystems such as forests and deserts will also be harmed.

The planet's climate could also be affected by depletion of the ozone layer. Wind patterns could change, resulting in climatic changes throughout the world.

SOLUTIONS

He discovery of the ozone depletion problem came as a great surprise. Now, action must be taken to ensure that the ozone layer is not destroyed. Because CFCs are so widespread and used in such a great variety of products, limiting their use is hard. Also, since many products already contain components that use CFCs, it would be difficult if not impossible to eliminate those CFCs already in existence.

The CFC problem may be hard to solve because there are already great quantities of CFCs in the environment. CFCs would remain in the stratosphere for another 100 years even if none were ever produced again.

Despite the difficulties, international action has been taken to limit CFCs. In the Montreal Protocol, 30 nations worldwide agreed to reduce usage of CFCs and encouraged other countries to do so as well. However, many environmentalists felt the treaty did "too little, too late", as theNatural Resources Defence Council put it. The treaty asked for CFC makers to only eliminate half of their CFC production, making some people feel it was inadequate.

By the year 2000, the US and twelve nations in Europe have agreed to ban all use and production of CFCs. This will be highly significant, because these countries produce three quarters of the CFCs in the world. Many other countries have signed treaties and written laws restricting the use of CFCs. Companies are finding substitutes for CFCs, and people in general are becoming more aware of the dangers of ozone depletion.

CHEMICAL PROCESSES

About 90 per cent of the Earth's protective ozone layer resides in the stratosphere between 15km and 50km altitude. Molecular oxygen is broken down in the stratosphere by solar radiation to yield atomic oxygen, which then combines with molecular oxygen to produce ozone. Ozone is destroyed naturally through a series of catalytic cycles involving oxygen, nitrogen, hydrogen and to a lesser extent chlorine and bromine species.

The abundance of stratospheric ozone is therefore chemically controlled by the stratospheric abundances of compounds containing hydrogen, nitrogen, chlorine and bromine. Increases in the abundances of methane and nitrous oxide (sources of hydrogen and nitrogen oxides respectively) thus affect the abundance and distribution of stratospheric ozone. Stratospheric ozone is also affected by the abundance of carbon dioxide (CO_2), because the rates of the chemical reactions that control the abundance of ozone are temperature-dependent, and the abundance of CO_2 plays a key role in determining the temperature structure of the stratosphere.

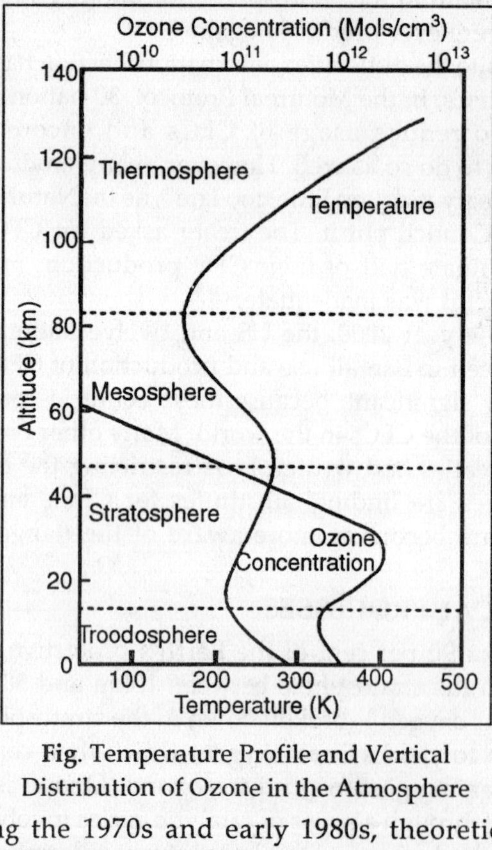

Fig. Temperature Profile and Vertical Distribution of Ozone in the Atmosphere

During the 1970s and early 1980s, theoretical model calculations focused on predicting the response of stratospheric ozone to changes in chlorine, assuming that the atmospheric abundances of other trace gases remained constant. However, since the early 1980s, with advances in understanding of trace gas trends and model formulation, model calculations have been used to predict the response of stratospheric ozone to simultaneous increases in chlorine (from chlorofluorocarbons, hydrochlorofluorocarbons, carbon tetrachloride, methylchloroform and methyl chloride) and bromine (from halons and methyl bromide), as well as methane, nitrous oxide and carbon dioxide.

RESPONSES

Policy Responses

Summarizes the main international policy responses in the 1970s and 1980s to the ozone layer issue. Established by UNEP in the 1970s, the Coordinating Committee on the ozone Layer (CCOL) has played a major role in establishing scientific research priorities, and in describing the implications of the latest scientific findings in terms understandable to policy-makers.

In 1980, for example, the CCOL reported that continued releases of CFCs would eventually deplete the ozone layer, and that this could have serious impacts on the health of people and the biosphere. Subsequently, the CCOL synthesized the knowledge base to provide the scientific basis for the Montreal Protocol. The policy implications of the scientific findings of the last decade have been profound, leading, for example, to a significant revision of the control measures of the Montreal Protocol.

The most recent scientific findings also have substantial policy implications and could lead to further significant revisions in the control measures of the Montreal Protocol at the Fourth meeting of the Contracting Parties.

Simulation Models in Exploring Policy

Nowadays, simulation models are used more and more to explore policy options for resolving the ozone depletion issue. A particular example, performed, was a sensitivity study of atmospheric chlorine-loading scenarios for a range of global emissions of compounds currently regulated under the Montreal Protocol, and for possible new HCFC substitutes. The results are summarized in Figure below, which shows that:

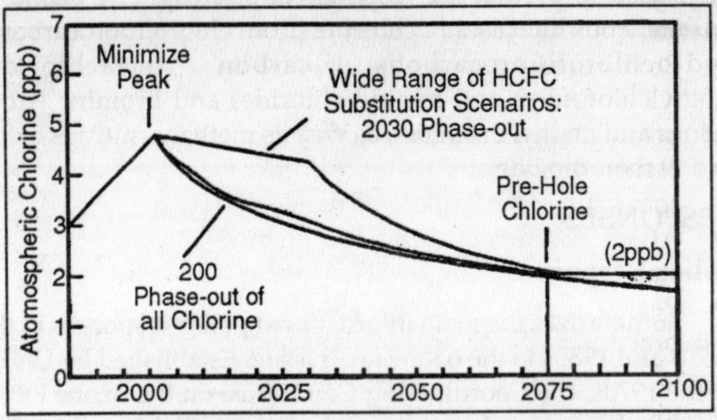

Fig. Impact of Chloroflurocarbon (CFC) and Hydrochlorofluorocarbon (HVFC) Substitution Policies on Atmospheric Abundance of Chlorine (CI).

- Peak chlorine loading is controlled by the date of the global phase-out (assumed to be the year 2000) for long-lived CFCs. and by emission rates between 1990 and 2000;
- Even with a complete phase-out of all chlorine emissions in 2000, the atmospheric level of chlorine will not decrease to 2 ppbv (the level of chlorine prior to the formation of the Antarctic ozone hole) until about 2075;
- a wide range of HCFC substitution policies can be implemented without delaying recovery of the Antarctic ozone hole, assuming that there is a phase-out of HCFC substitutes before the middle of the next century (the required phase-out date for the HCFCs is dependent upon their lifetime); and
- The additional atmospheric loading of chlorine due to HCFC substitution depends upon the HCFC emission rates and the lifetimes of the HCFCs, but is typically limited to less than 1 ppbv even with

significant initial substitution rates and subsequent growth in these emission rates.

The study concluded that in order to minimize future ozone layer depletion and to facilitate the most rapid elimination of the Antarctic ozone hole (i.e. by about 2075):

- Production and emission of long-lived halocarbons (CFCs and CCl_4) should be phased out as soon as possible, with the greatest possible global compliance:
- Halocarbon substitutes should be phased out sometime in the middle of the next century:
- Atmospheric concentrations of bromine should be stabilized at current levels or below.
- Peak chlorine loadings should be reduced, most practicably by significant cutbacks in the emissions of abundant, short-lived halocarbons, particularly methyl chloroform.

CONCLUDING REMARKS

Depletion of the stratospheric ozone layer by halogenated chemicals is a global problem. While the issue was primarily a matter of scientific curiosity during the 1970s and early 1980s, it has now become an urgent policy question for governments in both developed and developing countries. The weight of scientific evidence strongly indicates that chlorinated and brominated chemicals (largely man-made) are primarily responsible for the substantial decreases in stratospheric ozone (greater than 50 per cent in the total column) that occur over Antarctica every Southern Hemisphere springtime.

In addition, satellite data supported by ground-based observations have demonstrated that ozone has been decreasing since the late 1960s at middle and high latitudes in both the northern and southern hemispheres, and that this decrease cannot be explained by known natural processes. Indeed, the conclusion of the 1991 International Scientific Assessment was that 'the weight of evidence suggests that the observed middle- and high-latitude ozone losses are largely due to chlorine and bromine'.

The current and historic use of CFCs and other such

chemicals in developed countries is the primary cause of the problem. However, adequate protection of the ozone layer will require a full partnership between the developed countries, whose industries and consumers have caused the problem, and the developing countries, whose people wish to use these chemicals in refrigeration and other ways.

As in the case of atmospheric concentrations of lead, which are now declining, even in Greenland, as a result of tighter emission controls, the stratosphere would gradually recover if emissions of CFCs and similar chemicals were to cease. Thus, the most urgent task is to strengthen the Montreal Protocol and to develop an international method of monitoring national Programmemes to control CFCs and related gases. However scientific research and monitoring should continue into the next century to provide a basis for better policies.

GREENHOUSE EFFECT

The greenhouse effect is a naturally occurring process that aids in heating the Earth's surface and atmosphere. It results from the fact that certain atmospheric gases, such as carbon dioxide, *water vapour*, and methane, are able to change the energy balance of the planet by absorbing longwave radiation emitted from the Earth's surface. Without the greenhouse effect life on this planet would probably not exist as the average temperature of the Earth would be a chilly -18° Celsius, rather than the present 15° Celsius.

As energy from the Sun passes through the atmosphere a number of things take place. A portion of the energy (26% globally) is reflected or scattered back to space by clouds and other atmospheric particles. About 19% of the energy available is absorbed by clouds, gases (like ozone), and particles in the atmosphere. Of the remaining 55% of the solar energy passing through the Earth's atmosphere, 4% is reflected from the surface back to space. On average, about 51% of the Sun's radiation reaches the surface. This energy is then used in a number of processes, including the heating of the ground surface; the melting of ice and snow and the evaporation of water; and plant photosynthesis.

The heating of the ground by sunlight causes the Earth's surface to become a radiator of energy in the longwave band (sometimes called infrared radiation).

Absorption of longwave radiation by the atmosphere causes additional heat energy to be added to the Earth's atmospheric system. The now warmer atmospheric greenhouse gas molecules begin radiating longwave energy in all directions.

Over 90% of this emission of longwave energy is directed back to the Earth's surface where it once again is absorbed by the surface. The heating of the ground by the longwave radiation causes the ground surface to once again radiate, repeating the cycle described above, again and again, until no more longwave is available for absorption.

The amount of heat energy added to the atmosphere by the greenhouse effect is controlled by the concentration of greenhouse gases in the Earth's atmosphere. All of the major greenhouse gases have increased in concentration since the beginning of the Industrial Revolution (about 1700 AD). As a result of these higher concentrations, scientists predict that the greenhouse effect will be enhanced and the Earth's climate will become warmer.

Predicting the amount of warming is accomplished by computer modeling. Computer models suggest that a doubling of the concentration of the main greenhouse gas, carbon dioxide, may raise the average global temperature between 1 and 3° Celsius.

However, the numeric equations of computer models do not accurately simulate the effects of a number of possible negative feedbacks. For example, many of the models cannot properly simulate the negative effects that increased cloud cover would have on the radiation balance of a warmer Earth.

Increasing the Earth's temperature would cause the oceans to evaporate greater amounts of water, causing the atmosphere to become cloudier.

These extra clouds would then reflect a greater proportion of the Sun's energy back to space reducing the amount of solar radiation absorbed by the atmosphere and the Earth's surface.

With less solar energy being absorbed at the surface, the effects of an enhanced greenhouse effect may be counteracted.

Artificially created chlorofluorocarbons are the strongest greenhouse gas per molecule. However, low concentrations in the atmosphere reduce their overall importance in the enhancement of the greenhouse effect. Current measurements in the atmosphere indicate that the concentration of these chemicals may soon begin declining because of reduced emissions.

Reports of the development of ozone holes over the North and South Poles and a general decline in global stratospheric ozone levels over the last two decades has caused many nations to cutback on their production and use of these chemicals. In 1987, the signing of the Montreal Protocol agreement by forty-six nations established an immediate timetable for the global reduction of chlorofluorocarbons production and use.

Since 1750, methane concentrations in the atmosphere have increased by more than 150%. The primary sources for the additional methane added to the atmosphere (in order of importance) are rice cultivation, domestic grazing animals, termites, landfills, coal mining, and oil and gas extraction. Anaerobic conditions associated with rice paddy flooding results in the formation of methane gas. However, an accurate estimate of how much methane is being produced from rice paddies has been difficult to obtain. More than 60% of all rice paddies are found in India and China where scientific data concerning emission rates are unavailable.

Nevertheless, scientists believe that the contribution of rice paddies is large because this form of crop production has more than doubled since 1950. Grazing animals release methane to the environment as a result of herbaceous digestion. Some researchers believe the addition of methane from this source has more than quadrupled over the last century. Termites also release methane through similar processes. Land-use change in the tropics, due to deforestation, ranching, and farming, may be causing termite numbers to expand. If this assumption is correct, the contribution from these insects may be important.

Methane is also released from landfills, coal mines, and gas and oil drilling. Landfills produce methane as organic wastes decompose over time. Coal, oil, and natural gas deposits release methane to the atmosphere when these deposits are excavated or drilled.

The average concentration of nitrous oxide in the atmosphere is now increasing at a rate of 0.2 to 0.3% per year. Sources for this increase include land-use conversion; fossil fuel combustion; biomass burning; and soil fertilization. Most of the nitrous oxide added to the atmosphere each year comes from deforestation and the conversion of forest, savanna and grassland ecosystems into agricultural fields and rangeland.

Both of these processes reduce the amount of nitrogen stored in living vegetation and soil through the decomposition of organic matter. Nitrous oxide is also released into the atmosphere when fossil fuels and biomass are burned. However, the combined contribution of these sources to the increase of this gas in the atmosphere is thought to be minor. The use of nitrate and ammonium fertilizers to enhance plant growth is another source of nitrous oxide. Accurate measurements of how much nitrous oxide is being released from fertilization have been difficult to obtain. Estimates suggest that the contribution from this source may represent from 50% to 0.2% of nitrous oxide added to the atmosphere annually.

Ozone's role in the enhancement of the greenhouse effect has been difficult to determine scientifically. Accurate measurements of past long-term (more than 25 years in the past) levels of this gas in the atmosphere are currently unavailable. Concentrations of ozone gas are found in two different regions of the Earth's atmosphere. The majority of the ozone (about 97%) found in the atmosphere is localized in the stratosphere at an altitude of 15 to 55 Kilometres above the Earth's surface.

In recent years, the concentration of the stratospheric ozone has been decreasing because of the buildup of chlorofluorocarbons in the atmosphere. Since the late 1970s, scientists have discovered that total column ozone amounts over Antarctica in the springtime have decreased by as much as 70%. Satellite measurements have indicated that the zone from 65° North to 65° South latitude has had a 3% decrease in stratospheric ozone since 1978. Ozone is also highly concentrated at the Earth's surface. Most of this ozone is created as an artificial by product of photochemical smog.

In summary, the greenhouse effect causes the atmosphere to trap more heat energy at the Earth's surface and within the atmosphere by absorbing and re-emitting longwave energy. Of the longwave energy emitted back to space, 90% is intercepted and absorbed by greenhouse gases. Without the greenhouse effect the Earth's average global temperature would be -18° Celsius, rather than the present 15° Celsius.

In the last few centuries, the activities of humans have directly or indirectly caused the concentration of the major greenhouse gases to increase.

Scientists predict that this increase mayenhance the greenhouse effect making the planet warmer. Some experts estimate that the Earth's average global temperature has already increased by 0.3 to 0.6° Celsius, since the beginning of this century, because of this enhancement. Predictions of future climates indicate that by the middle of the next century the Earth's global temperature may be 1 to 3° Celsius higher than today.

BASIC MECHANISM

The Earth receives energy from the Sun mostly in the form of visible light. The bulk of this energy is not absorbed by the atmosphere since the atmosphere is transparent to visible light. 50% of the sun's energy reaches the Earth which is absorbed by the surface as heat. Because of its temperature, the Earth's surface radiates energy in infrared range.

The Greenhouse gases are not transparent to infrared radiation so they absorb infrared radiation. Infrared radiation is absorbed from all directions and is passed as heat to all gases in the atmosphere. The atmosphere also radiates in the infrared range (because of its temperature, in the same way the Earth's surface does) and does so in all directions. The surface and lower atmosphere are warmed because of the greenhouse gases and makes our life on earth possible.

DETAILED EXPLANATION

The Earth receives energy from the Sun in the form of radiation. Most of the energy is in visible wavelengths and in infrared wavelengths that are near the visible range (often called "near infrared"). The Earth reflects about 30% of the incoming solar radiation. The remaining 70% is absorbed, warming the land, atmosphere and ocean.

For the Earth's temperature to be in steady state so that the Earth does not rapidly heat or cool, this absorbed solar radiation must be very closely balanced by energy radiated back to space in the infrared wavelengths. Since the intensity of infrared radiation increases with increasing temperature, one can think of the Earth's temperature as being determined by the infrared flux needed to balance the absorbed solar flux. The visible solar radiation mostly heats the surface, not the atmosphere, whereas most of the infrared radiation escaping to space is emitted from the upper atmosphere, not the surface. The infrared photons emitted by the surface are mostly absorbed in the atmosphere by greenhouse gases and clouds and do not escape directly to space.

The reason this warms the surface is most easily understood by starting with a simplified model of a purely radiative greenhouse effect that ignores energy transfer in the atmosphere by convection (sensible heat transport, Sensible heat flux) and by the evaporation and condensation of water vapour (latent heat transport, Latent heat flux). In this purely radiative case, one can think of the atmosphere as emitting infrared radiation both upwards and downwards. The upward infrared flux emitted by the surface must balance not only the absorbed solar flux but also this downward infrared flux emitted by the atmosphere. The surface temperature will rise until it generates thermal radiation equivalent to the sum of the incoming solar and infrared radiation.

A more realistic picture taking into account the convective and latent heat fluxes is somewhat more complex. But the following simple model captures the essence. The starting point is to note that the opacity of the atmosphere to infrared radiation determines the height in the atmosphere from which most of the photons are emitted into space. If the atmosphere is more opaque, the typical photon escaping to space will be emitted from higher in the atmosphere, because one then has to go to higher altitudes to see out to space in the infrared. Since the emission of infrared radiation is a function of temperature, it is the temperature of the atmosphere at this emission level that is effectively determined by the requirement that the emitted flux balance the absorbed solar flux.

But the temperature of the atmosphere generally decreases with height above the surface, at a rate of roughly 6.5 °C per Kilometre on average, until one reaches the stratosphere 10 15 km above the surface. (Most infrared photons escaping to space are emitted by the troposphere, the region bounded by the surface and the stratosphere, so we can ignore the stratosphere in this simple picture.) A very simple model, but one that proves to be remarkably useful, involves the assumption that this temperature profile is simply fixed, by the non-radiative energy fluxes.

Given the temperature at the emission level of the infrared flux escaping to space, one then computes the surface temperature by increasing temperature at the rate of 6.5 °C per Kilometre, the environmental lapse rate, until one reaches the surface. The more opaque the atmosphere, and the higher the emission level of the escaping infrared radiation, the warmer the surface, since one then needs to follow this lapse rate over a larger distance in the vertical. While less intuitive than the purely radiative greenhouse effect, this less familiar radiative-convective picture is the starting point for most discussions of the greenhouse effect in the climate modeling literature.

ACID RAIN

DEFINITION

Acid rain is rainfall that has been acidified. Rain is

acidified by oxides of sulfur and nitrogen. Acid rain usually has a pH of less than 5.6. Acid rain is formed when pollutants called oxides of sulfur and nitrogen, contained in power plant smoke, factory smoke, and car exhaust, react with the moisture in the atmosphere. Dry deposition, such as soot and ash, sleet, hail, snow, smog and low level ozone are forms that acid rain can take, despite its name.

KEY CONCEPTS

Rain from an unpolluted atmosphere has a pH close to 6.0 (slightly acidic).

This acidity is due to the reaction of water vapour and non-metal oxides in the atmosphere, such as carbon dioxide and nitrogen oxide, forming dilute acids.

- *Carbon dioxide reacts with water to form carbonic acid*:
$$CO_{2(g)} + H_2O_{(l)} = H_2CO_{3(aq)}$$
- *Since carbonic acid is a weak acid it partially dissociates*:
$$CO_{2(g)} + H_2O_{(l)} = H^+_{(aq)} + HCO_3^-{}_{(aq)}$$
- Nitrogen dioxide reacts with water to form a mixture of nitrous acid and nitric acid:
$$2NO_{2(g)} + H_2O_{(l)} = HNO_{2(aq)} + HNO_{3(aq)}$$

Acid rain has a pH below 5.6 due mainly to the reaction of water vapour with sulfur dioxide and the oxides of nitrogen.

- Sulfur dioxide reacts with water to form sulfurous acid (H_2SO_3):
$$SO_2(g) + H_2O(l) = H_2SO_3(aq)$$
- Sulfur dioxide (SO_2) can be oxidised gradually to sulfur trioxide (SO_3):
$$2SO_2(g) + O_2(g) \rightarrow 2SO_3(g)$$

- Sulfur trioxide (SO_3) reacts with water to form sulfuric acid (H_2SO_4):

$$SO_3(g) + H_2O(l) \rightarrow H_2SO_4(aq)$$

- Oxides of nitrogen, particularly nitrogen dioxide (NO_2) react with water to form nitrous acid (HNO_2) and nitric acid (HNO_3):

$$2NO_2(g) + H_2O(l) \rightarrow HNO_2(aq) + HNO_3(aq)$$

Table. Sources of the Acids in Clean and Polluted Air

Acid	Carbonic acid (H_2CO_3)
Clean air	Natural carbon dioxide (CO2) produced during plant and animal respiration produced during decomposition of organic matter
Polluted air	Carbon dioxide (CO2) released from the combustion of fuels The complete combustion of coal: $C(s) + O_2(g)\ CO_2(g)$ The complete combustion of petrol (for example octane, C8H18(l)): 2C8H18(l)+25O2(g)--->16CO2(g)+18H2O(g) Complete combustion of ethanol (ethyl alcohol): $C_2H5OH(l)+3O_2(g) \rightarrow 2CO_2(g)+3H_2O(g)$
Acid	Formic acid (Methanoic acid) (HCOOH)
Clean air	Oxidation of natural methane (CH4) formed during the anaerobic decomposition of organic matter
Polluted air	Increased oxidation
Acid	Sulfuric acid (Sulphuric acid) (H_2SO_4)
Clean air	Natural decay of organic matter releases hydrogen sulfide gas (H2S) which can be oxidised to sulfur dioxide (SO2): $2H_2S(g)+3O_2(g) \rightarrow 2SO_2(g)+2H_2O(l)$ Sulfur dioxide can be oxidised to sulfur trioxide (SO3): $2SO_2(g)+O_2(g) \rightarrow 2SO_3(g)$ Sulfur trioxide then reacts with water to form sulfuric acid: $SO_3(g)+H_2O(l) \rightarrow H_2SO_4(aq)$ Volcanoes emit sulfur dioxide

	which can be oxidised to sulfur trioxide which then reacts with water forming sulfuric acid. Ocean algae release sulfur gases such as dimethyl sulfide which is oxidised to form sulfuric acid.
Polluted air	Combustion of coal and other fossil fuels account for about 80% of the man-made sulfur dioxide in the atmosphere (sulfur is present in the proteins of the original living matter that has been fossilised to produce the fossil fuel such as coal, oil or petroleum). Most of this is from coal-fired power stations, motor vehicle emmissions account for only about 1% of the sulfur dioxide present. Sulfur dioxide is also produced when sulfur ores are roasted: $2ZnS(s)+3O2(g)\rightarrow 2ZnO(s)+2SO2(g)$ Sulfur dioxide is also produced in the manufacture of sulfuric acid by the contact process, in petroleum refining and in the manufacture of coke from coal
Acid	Nitric acid (HNO_3)
Clean air	Lightning flashes lead to a reaction Between atmospheric nitrogen and oxygen in the presence of water vapour which forms nitric acid
Polluted air	Combustion of fossil fuels Nitric oxide (nitrogen monoxide, NO) is produced in internal combustion engines as a result of the reaction between oxygen and nitrogen at high temperatures: $N_2(g)+O_2(g)\ 2NO(g)$ Nitrogen monoxide is readily oxidised to nitrogen dioxide (NO2): $2NO(g)+O_2(g)\ 2NO_2(g)$ Nitrogen dioxide reacts with water to form nitrous acid (HNO_2) and nitric acid (HNO_3): $2NO_2(g)+H_2O(l)\ HNO_2(aq)+HNO_3(aq)$

Acid	Methanesulfonic acid (Methanesulphonic acid)
Clean air	Ocean algae emit dimethyl sulfide which oxidises in air to produce methanesulfonic acid.
Polluted air	Methanesulfonic acid is only produced naturally

EFFECTS OF ACID RAIN

Acid rain is a form of environmental pollution that damages buildings and marble statues by reacting with the calcium carbonate to form soluble calcium hydrogen carbonate (calcium bicarbonate, $Ca(HCO_3)_2$)

$$CaCO_3 + \text{acid rain} \rightarrow Ca(HCO_3)_{2(aq)}$$

Acid rain can leach aluminium from the soil into ground water, lakes and rivers, poisoning fish and plant roots. The sulfates and hydrogen sulfates in acid rain can can leach essential plant nutrients such as calcium and magnesium, from the soil. Acid rain disrupts the process of photosynthesis resulting in damage to plant life.

At low concentrations it retards the production of chlorophyll and at high concentrations it forms sulfuric acid which kills the plant. Some organisms are sensitive to changes of acidity in water which can affect their ability to reproduce and in some cases may kill them.

CAUSES

Acid rain can be caused by many things. Industrial emissions from factories and power plants that burn fuels such as natural gas, coal or oil, emit smoke that gives off oxides of sulfur and nitrogen is one cause. Another cause is vehicles (e.g. cars, buses) that burn gasoline and diesel. The exhaust emitted by burning these fuels contains sulfur dioxide, an oxide of sulfur. Also, vehicles that have gas engines will produce oxides of nitrogen, another cause of acid rain. One other cause is home fires giving off smoke that contains sulfur dioxide. Some fairly minor causes are natural causes which are volcanoes, swamps and rotting plants giving off sulfur dioxide. Natural causes only account for 10% of the pollution causing acid rain.

EFFECTS AND PROBLEMS

There are many problems and effects caused by acid rain. Acid rain can cause buildings, statues and bridges to deteriorate faster than usual. Another problem is it harms thousands of lakes, rivers, and streams worldwide. It disrupts lake ecosystems and kill wildlife in affected lakes, rivers and streams. Acid rain also damages soil and the tree roots in it. When soil is acidified, tree roots are damaged, leaving them not able to draw in enough nutrients to support the tree. When acid rain falls on trees, it makes their leaves turn brownish-yellow and the tree can no longer carry out photosynthesis properly. Another problem is it will harm people when they breathe in smog, acid rain in one of its many forms. Acid rain can also harm people indirectly. This happens when people eat fish caught in affected lakes or rivers. Also, if the water source is acidic enough, it will react with copper or lead pipes to harm humans. It also washes aluminum into the water supply. Birds can be harmed if they live in affected waters or feed on fish living in affected waters.

AREAS AFFECTED

There are several places around the world affected by acid rain and here are the main ones. The Northeastern section of the United States where acid rain is caused by high numbers of factories and power plants is one affected area. Also in that same region, the Southeastern section of Canada is affected and the main cause is factories in the Toronto-Hamilton area. Central Europe and Scandanavia (Sweden, Norway and Finland) are also affected, here being the British and other European factories doing the damage. One more main area that is affected is parts of Asia, specifically India and China, where acid rain is caused by large numbers of factories.

THE HISTORY OF THE ACID RAIN PROBLEM

Here is a little bit on the history of the acid rain problem.

The problem of acid rain probably originated during the 1730's, at the height of the industrial revolution. It was discovered in the 1950's and started being noticed in the 1960's. Since the 1960's, the problem has gotten worse in rural areas because the tall chimneys on factories allow the wind to transport pollutants far away from their sources. In 1984 it was reported that almost half of the trees in the famous black forest in Germany had been damaged by acid rain. In 1988, as part of the United Nations-sponsored Long Range Transboundary Air Pollution Agreement, the U.S.A. and 24 other nations agreed to some rules limiting nitrogen oxide emissions to 1987 levels.1990 changes to the Clean Air Act set rules to cut down the release of sulfur dioxide from power plants down to 10 million tonnes by January 1, 2000.

WHAT ACID RAIN HAS DONE

There are many bad things that have happened because of acid rain and here are some. In Sweden, drinking water once contained enough water to make people's hair turn green. In the Czech Republic, many trees lost all their leaves as a result of acid rain. The Taj Mahal in India, one of the ten wonders of the world, is being constantly threatened by acid rain. Some famous statues, such as the Lincoln Memorial and Michaelangelo's statue of Marcus Aurelius, have started deteriorating because of acid rain. In London in 1952, very thick acid smog killed 4,000 people.

POSSIBLE SOLUTIONS

There are many possible solutions to the acid rain problem. One is to use fuels that burn more cleanly. Another solution is to burn coal more efficiently. Power plant and factory chimneys can be fitted with scrubbers that release 90-95% sulfur free smoke and also produces sludge from which gypsum, a building material, can be produced. This problem can also be solved by enforcing tight emission standards on vehicles that burn gas and diesel. One other solution is fitting devices such as a catalytic converter into car exhaust pipes to minimize the amount of sulfur dioxide in car exhaust.

LA NINO

The term El Niño (Spanish for "the little boy" or "the Christ Child") was originally used by fishermen along the coasts of Ecuador and Peru to refer to above-normal sea-surface temperatures that typically appear around Christmas time in the eastern Pacific Ocean and last for several months. But El Niño's effects are not limited to Peru and Ecuador. They can impact weather patterns around the world, and the disruption of the normal climate can have profound and even tragic consequences.

WORKING OF EL NIÑO

In normal years, winds tend to blow from east to west across the waters of the tropical Pacific. The easterly winds (remember that wind directions refer to the direction FROM which the wind comes: an easterly trade wind comes from the EAST and blows WEST) push the surface waters westward across the ocean. In turn, this causes deeper, colder waters to rise to the surface. This "upwelling" of deep ocean waters brings with it nutrients that otherwise would remain near the bottom. Fish populations living in the upper waters are dependent on these nutrients for survival.

During El Niño years, the winds weaken, reducing or even choking off the upwelling of deep water. The consequent warming of the ocean surface further weakens the winds and strengthens El Niño. As the Pacific continues to heat up, the warmer waters shift eastward, and so do the clouds and thunderstorms that produce heavy rainfall along the equator. This results in changes in jet streams (the winds high aloft), which lead to dry conditions in Indonesia and Australia, and floods in Peru and Ecuador. El Niño events occur on average every 3 to 5 years, though there can be periods of up to a decade without an El Niño.

EL NIÑO'S EFFECTS

The 1982-83 El Niño was unusually strong. In Ecuador and northern Peru, up to 100 inches of rain fell during a six-month period, transforming the coastal desert into grassland dotted with lakes. Abnormal wind patterns also caused the monsoon rains to fall over the central Pacific instead of on the western shores, which led to droughts and disastrous forest fires in Indonesia and Australia. Overall, the loss to the global economy as a result of the El Niño amounted to more than $8 billion.

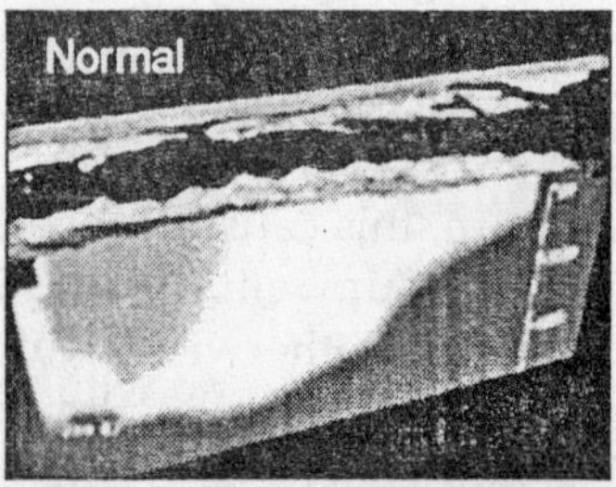

Likewise, the winter of 1997-1998 was marked by a record-breaking El Niño event. The result was unusual weather in parts of the world, including the U.S. Severe weather events included flooding in the southeastern United States, major storms in the Northeast, and flooding in California.

Chapter 5

Environmental Microbiology

DEFINITION

The study of how microorganisms affect the earth and its atmosphere is called environmental microbiology or microbial ecology. Environmental microbiology is the study of microbes in the environment and their interactions with each other. Microbes are the tiniest creatures on Earth, yet despite their small size, they have a have a huge impact on us and on our environment.

Most types of microbes remain unknown. It is estimated that we know fewer than 1% of the microbial species on Earth. Yet microbes surround us everywhere — air, water, soil. An average gram of soil contains one billion (1,000,000,000) microbes representing probably several thousand species. Environmental microbiology is the study of the relationships that exist between microorganisms and the environment.

Environmental microbiology is the relationship of microorganisms with themselves and with their surroundings. Microorganisms have special impact on the whole biosphere. They are the backbone of ecosystems of the zones where light cannot approach. In such zones, chemosynthetic bacteria are present which provide energy and carbon to the other organisms there. Some microbes are decomposers which have ability to recycle the nutrients. So, microbes have a special role in biogeochemical cycles. Microbes, especially bacteria, are of great importance in the sense that their symbiotic relationship (either positive or negative) have special effects on the ecosystem.

Bacteria and their enzymes, along with some fungi and critical nutrient additives are cost effective agents for in-situ remediation of hazardous wastes and subsurface pollution in soils, sediments and wastewaters. The ability of each bacteria strain to degrade toxic waste depends on the nature of each contaminant. Since most sites are typically comprised of multiple pollutant types, the most effective approach to bioremediation is to use a mixture of bacterial species/strains, each specific to the degradation of one or more types of contaminants. It is critical to monitor the composition of the indigenous and added bacterial consortium in order to evaluate the activity level of the bacteria, and to permit modifications of the nutrients and other conditions for optimizing the bioremediation process.

OVERVIEW

The pervasive and manifold positive and negative impacts of microbial activities on human and environmental health, and human activities, require that efforts to improve the human condition include efforts to modulate microbial activities. Attainment of this goal will be greatly facilitated by acquisition of a fundamental understanding of microbial functioning in their natural settings, and of parameters that regulate such activities. Our current fragmentary knowledge of microbial processes is based largely on information obtained from individual strains of microbes studied in the Labouratory as pure, exponentially-growing cultures in nutrient-rich, homogeneous aqueous solutions.

Sewage treatment is the process that removes the majority of the contaminants from waste-water or sewage and produces both a liquid effluent suitable for disposal to the natural environment and a sludge. To be effective, sewage must be conveyed to a treatment plant by appropriate pipes and infrastructure and the process itself must be subject to regulation and controls.

Sewage is the liquid waste from toilets, baths, showers, kitchens, etc. that is disposed of via sewers. In many areas sewage also includes some liquid waste from industry and commerce. In the UK, the waste from toilets is termed foul waste, the waste from items such as basins, baths, kitchens is termed sullage water, and the industrial and commercial waste is termed trade waste.

The division of household water drains into Greywater and Black water is becoming more common in the United States, with greywater being permitted to be used for watering plants or recycled for flushing toilets. Much sewage also includes some surface water from roofs or hard-standing areas. Municipal wastewater therefore includes residential, commercial, and industrial liquid waste discharges, and may include stormwater runoff.

Sewerage systems that transport liquid waste discharges and stormwater together to a common treatment facility are called combined sewer systems. The construction of combined sewers is a less common practice in the U.S. and Canada than in the past and is no longer accepted within Building Regulations in the UK. Instead, liquid waste and stormwater are collected and conveyed in separate sewer systems, referred to as sanitary sewers and storm sewers in the U.S. and as foul sewers and surface water sewers in the UK. Overflows from foul sewers designed to relieve pressure from heavy rainfall are termed storm sewers or combined sewer overflows.

As rainfall runs over the surface of roofs and the ground, it may pick up various contaminants including soil particles (sediment), metals, organic compounds, animal waste, and oil and grease. Some jurisdictions, such as certain communities located in southern California, require stormwater to receive some level of treatment before being discharged to the environment. Examples of treatment processes used for stormwater include sedimentation basins, wetlands, and vortex separators (to remove coarse solids).

"Conventional wastewater treatment" is the process used in modern wastewater treatment plants that removes the majority of the contaminants from wastewater. This produces a liquid effluent suitable for disposal to the natural environment and also produces a sludge.

The treatment process typically involves the following

three stages: Primary treatment Primary treatment is to reduce oils, grease, fats, sand, grit, and coarse (settleable) solids.

Grit removal: This stage typically includes a grit channel where the velocity of the incoming wastewater is carefully controlled to allow grit and stones to settle but still maintain all organic material within the flow. Grit and stones need to be removed early in the process to avoid damage to pumps and equipment in the remaining treatment stages.

Screening or maceration: The grit free liquid is then passed through fixed or rotating screens to remove larger material such as rags. Screenings are collected and may be returned to the sludge treatment plant or may be disposed of off site by landfilling or incineration. Maceration, in which solids are cut into small particles through the use of rotating knife edges mounted on a revolving cylinder, is used in plants that are able to process this particulate waste. Macerators are, however, more expensive to maintain and are less reliable than physical screens.

SEDIMENTATION

Primary sedimentation tank at a rural treatment plantIn almost all plants there is a sedimentation stage where the sewage is allowed to stand in large tanks so that faecal solids can settle and floating material such as grease and plastics can rise to the surface and be skimmed off. The main purpose of the primary stage is to produce a generally homogeneous liquid capable of being treated biologically together with a sludge that can be separately treated or processed. Primary settlement tanks are usually equipped with mechanically driven scrapers that continually drive the collected sludge towards a hopper in the base of the tank from where it can be pumped to further sludge treatment stages.

Secondary treatment: Secondary treatment is designed to substantially degrade the biological content of the sewage. The majority of municipal and industrial plants treat the settled sewage liquor using aerobic biological processes. For this to be effective, the biota require both oxygen and a substrate on which to live. There are number of ways in which this is done.

In all these methods, the bacteria and protozoa consume biodegradable soluble organic contaminants (e.g. sugars, fats, organic short-chain carbon molecules, etc.) and bind much of the less soluble fractions into floc particles.

Roughing Filters: Roughing filters are intended to treat particularly strong or variable organic loads. They are typically tall, columnar filters filled with open synthetic filter media to which sewage is applied at a relatively high rate. The design of the filters allows high hydraulic loading and a high flow-through of air. The resultant liquor is usually within the normal range for conventional treatment processes.

Activated sludge: Activated sludge plants use a variety of mechanisms and processes to use dissolved oxygen to generate a biological floc that substantially removes organic material. It also traps particulate material and can, under ideal conditions, convert ammonia to nitrite or nitrate.

FILTER BEDS

Trickling filter bed using plastic mediaIn older plants and plants receiving more variable loads, trickling filter beds are used where the settled sewage liquor is spread onto the surface of a deep bed made up of coke (carbonised coal), rocks or specially fabricated plastic media with high surface areas. The liquor is distributed through perforated rotating arms radiating from a central pivot.

The distributed liquor trickles through this bed and is collected in drains at the base. These drains also provide a source of air which percolates up through the bed, keeping it aerobic. Biological film comprising of bacteria, protozoa and fungi forms on all the available surfaces and this provides the required biological treatment capability to effect the reduction in organic content.

Rotating Plates and Spirals In some smaller plants slowly revolving plates or spirals are used which are partially submerged in the liquor. A biotic floc is created which provides the required substrate.

Secondary sedimentation: The final step in the secondary treatment stage is to settle out the biological floc or filter material and produce an effluent with very low levels of organic material and suspended matter.

Secondary Sedimentation tank at a rural treatment plantTertiary treatment Tertiary treatment provides a final stage to raise the effluent quality to the standard required before it is discharged to the receiving environment (sea, river, lake, ground, etc.) More than one tertiary treatment process may be used at any treament plant. If disinfection is practiced, it is always the final process.

Effluent polishing Sand filtration Sand filtration removes much of the residual suspended matter. Lagooning Lagoons provides settlement and further biological improvement through storage in large man-made ponds or lagoons.

Constructed wetlands These include engineered reed beds and a range of similar methodologies, all of which provide a high degree of aerobic biological improvement and can often be used instead of secondary treatment for small communities.

Nutrient removal Wastewater may also contain high levels of nutrients (nitrogen and phosphorus) that in certain forms may be toxic to fish and invertebrates at very low concentrations(e.g. ammonia) or that can create nuisance conditions in the receiving environment (e.g. weed or algal growth). Although the growth of weeds and algae may seem to be primarily an aesthetic issue, algae can produce toxins, and in dying their decay and consumption by bacteria in the environment can result in the depletion of oxygen in the water and the possible consequential suffocation of fish.

Where receiving rivers discharge to lakes or shallow seas, the added nutrients can cause severe and sometimes irreversible eutrophication with the loss of many sensitive clean water species. The removal of nitrogen and/or phosphorus from wastewater can be achieved either biologically or by chemical precipitation treatment processes.

Nitrogen removal Biological treatment of nitrogen generally involves creating conditions within the treatment process for bacteria to convert the ammonia to nitrate, and then allowing other bacteria to reducing the nitrate to nitrogen gas, which is released to the atmosphere. Sand filters, lagooning and the use of reed beds can all be used to reduce nitrogen. Sometimes the conversion of toxic ammonia to nitrate alone is referred to as tertiary treatment.

Phosphorous removal The biological treatment of wastewater to remove phosphorus also involves the design and creation of specific environmental conditions within a treatment plant to enable specific bacteria to bio-accumulate large quantities of phosphorus. When the bacteria containing the phosphorus are removed, the resulting bacterial biosolids often have a high fertilizer value. Phosphorus can also be removed by chemical precipitation using (commonly) salts of iron (i.e. ferric chloride) or aluminum (i.e. alum). The resulting chemical sludge, however, is difficult to dispose of, and the use of chemicals in the treatment process is expensive and makes operation difficult and often messy.

Disinfection Disinfecting substantially reduces the numbers of living organisms in the water to be discharged. The effectiveness of disinfection depends on the quality of the water being treated, the type of disinfection being used, the application rate, the contact time and environmental variables. Turbid water will be treated less successfully since solid matter can shield organisms, especially from Ultraviolet light or if contact times are low. Generally, short contact times, low doses and high flows all mitigate against effective disinfection. Common methods of disinfection include ozone, chlorine, or UV light. Chloramine, which is used for drinking water, is not used in waste water treatment because of its persistence.

A highly efficient dye-deColourizing bacterium, strain S12(T), was isolated from activated sludge of a textile-printing waste-water treatment plant in Guangzhou, China. The cells were Gram-negative and motile by means of a single polar flagellum. The strain was capable of anaerobic growth either by fermentation of glucose or by anaerobic respiration and utilized a variety of electron acceptors, including nitrate, iron oxide and thiosulfate. The physiological properties, tested by using the Biolog GN2 system, were similar to those of the genus of Shewanella. Analysis of the nearly complete 16S rRNA gene sequence of strain S12(T) showed the highest similarity (98 and 97 %, respectively) to Shewanella baltica and Shewanella putrefaciens.

However, the level of gyrB similarity between strain S12(T) and S. putrefaciens was 87 %. DNA from strain S12(T) showed 41.8 and 41.9 % DNA relatedness, respectively, to the DNA of S. baltica DSM 9439(T) and S. putrefaciens DSM 6067(T). The DNA G+C content of strain S12(T) was 49.3 mol%. The predominant menaquinone was MK-7 and the predominant ubiquinones were Q-7 and Q-8. The dominant fatty acids were 15: 0, 16: 0, iso-15: 0 and 16: 1omega7c, similar to the profiles of other Shewanella species.

On the basis of its physiological and molecular properties, strain S12(T) appears to represent a novel species of the genus Shewanella, for which the name Shewanella deColourationis sp. nov. is proposed. The type strain is S12(T) (=CCTCC M 203093(T)=IAM 15094(T)).

Hydrogenophaga palleronii (98.5 %) and Hydrogenophaga taeniospiralis (98.0 %). The fatty acid patterns and substrate-utilization profiles displayed similarity to the those of the five Hydrogenophaga species with validly published names, although clear differentiating characteristics were also observed. The two strains showed DNA-DNA hybridization values of 51 % with respect to each other. No close similarities to any other Hydrogenophaga species were detected in hybridization experiments with the genomic DNAs.

On the basis of these results, two novel Hydrogenophaga species, Hydrogenophaga defluvii sp. nov. and Hydrogenophaga atypica sp. nov. are proposed, with BSB 9.5(T) (=DSM 15341(T)=CIP 108119(T)) and BSB 41.8(T) (=DSM 15342(T)=CIP 108118(T)) as the respective type strains.

Effect of ethoxylate number and alkyl chain length on the pathway and kinetics of linear alcohol ethoxylate biodegradation in activated sludge; Itrich NR et al.; Batch activated-sludge die-away studies were conducted with various pure homologs to determine the effect of ethoxylate number and alkyl chain length on the kinetics of primary and ultimate biodegradation of linear alcohol ethoxylates.

The 14C-(ethoxylate) homologs C14E1, C14E3, C14E6, and

C14E9 were used to investigate the effect of ethoxylate number, and 14C-(ethoxylate) homologs C12E6, C14E6, and C16E6 were used to examine the effect of chain length. Activated sludge was dosed with a trace concentration (0.2 microM) of each homolog, and the disappearance of parent, formation of metabolites, production of 14CO2, and uptake into solids were monitored with time. Ethoxylate number had little effect on the first-order decay rates for primary biodegradation, which ranged from 61 to 78 h(-1).

However, alkyl chain length had a larger effect, with the C16 chain-length homolog exhibiting a slower rate of parent decay (18 h(-1)) compared to its corresponding C12 and C14 homologs (61-69 h(-1)). Ethoxylate number affected the mechanism of biodegradation, with fission of the central ether bond to yield the corresponding fatty alcohol and (poly)ethylene glycol group increasing in dominance with increasing ethoxylate number. Based upon the measured rates of primary biodegradation, removal of parent during activated-sludge treatment was predicted to range between 99.7 and 99.8% for all homologs except C16E6, which had a predicted removal of 98.9%.

Based upon the measured rates of ultimate biodegradation, removal of ethoxylate-containing metabolites was predicted to exceed 83% for all homologs. These predictions corresponded closely with previously published removal measurements in Labouratory continuous activated-sludge systems and actual treatment plants.

It was found that the estimated theoretical maximum growth yield of aerobic granules was as low as 0.2 g biomass g(-1) chemical oxygen demand (COD) and 10-40% of input substrate-COD was consumed through the maintenance metabolism, while experimental results further showed that the unit oxygen uptake by aerobic granules was 0.68 g oxygen g(−1) COD, which was much higher than that reported in activated sludge processes. Based on the growth yield and unit oxygen uptake determined, an oxidative assimilation equation of acetate-fed aerobic granules was derived; and this was confirmed by respirometric tests. In aerobic granular culture, about 74% of the input substrate-carbon was converted to carbon dioxide. The growth yield of aerobic granules was three times lower than that of activated sludge. It is likely that high carbon dioxide production is the main cause of the low growth yield of aerobic granules, indicating a possible energy uncoupling in aerobic

GRANULAR CULTURE

Comparing activated sludge and aerobic granules as microbial inocula for phenol biodegradation; Tay ST et al.; Activated sludge and acetate-fed granules were used as microbial inocula to start up two sequencing batch reactors (R_1, R_2) for phenol biodegradation. The reactors were operated in 4-h cycles at a phenol loading of 1.8 kg m(–3) day(–1). The biomass in R1 failed to remove phenol and completely washed out after 4 days. R_2 experienced initial difficulty in removing phenol, but the biomass acclimated quickly and effluent phenol concentrations declined to 0.3 mg l(–1) from day 3.

The acetate-fed granules were covered with bacterial rods, but filamentous bacteria with sheaths, presumably to shield against toxicity, quickly emerged as the dominant morphotype upon phenol exposure. Bacterial adaptation to phenol also took the form of modifications in enzyme activity and increased production of extracellular polymers. 16S rRNA gene fingerprints revealed a slight decrease in bacterial diversity from day 0 to day 3 in R1, prior to process failure.

In R_2, a clear shift in community structure was observed as the seed evolved into phenol-degrading granules without losing species-richness. The results highlight the effectiveness of granules over activated sludge as seed for reactors treating toxic wastewaters.

Processes for managing pathogens; Godfree A et al.; Wastewater contains human, animal, and plant pathogens capable of causing viral, bacterial, or parasitic infections. There are several routes whereby sewage pathogens may affect human health, including direct contact, contamination of food crops, zoonoses, and vectors. The range and numbers of pathogens in municipal wastewater vary with the level of endemic disease in the community, discharges from commercial activities, and seasonal factors.

Regulations to control pathogen risk in the United States and Europe arising from land application of biosolids are based on the concept of multiple barriers to the prevention of transmission.

The barriers are:

- Treatment to reduce pathogen content and vector attraction.
- Restrictions on crops grown on land to which biosolids have been applied.
- Minimum intervals following application and grazing or harvesting.

Wastewater treatment reduces number of pathogens in the wastewater by concentrating them with the solids in the sludge. Although some treatment processes are designed specifically to inactivate pathogens, many are not, and the actual mechanisms of microbial inactivation are not fully understood for all processes. Vector attraction is reduced by stabilization (reduction of readily biodegradable material) and/or incorporation immediately following application.

Concerns about health risks have renewed interest in the effects of treatment (on pathogens) and advanced treatment methods, and work performed in the United States suggests that Class A pathogen reduction can be achieved less expensively than previously thought. Effective pathogen risk management requires control to the complete chain of sludge treatment, biosolids handling and application, and post-application activities. This may be achieved by adherence to quality management systems based on hazard analysis critical control point (HACCP) principles.

Evaluation of microscopic techniques (epifluorescence microscopy, CLSM, TPE-LSM) as a basis for the quantitative image analysis of activated sludge; Lopez C et al.; Microscopic techniques ranging from epifluorescence microscopy to confocal laser scanning microscopy (CLSM) and two photon excitation laser scanning microscopy (TPE-LSM) combined with fluorescent stains can help to evaluate complex microbial aggregates such as activated sludge flocs. To determine the application limits of these microscopic techniques, activated sludge samples from three different sources were evaluated after staining with a fluorescent viability indicator (Baclight Bacterial Viability Kit, Molecular Probes).

Image analysis routines were developed to quantify overall amounts of red and green stained cells, location of stained cells within the flocs, and the spatial organization in clusters and filaments. It was found that the selection of the appropriate microscopic technique depends strongly on the type of microbial aggregates being Analysed.

For flocs with high cell density, the use of TPE-LSM is preferred, since it provides a clearer image of the internal structure of the aggregate. Epifluorescence microscopy did not allow to reliably quantify red stained cells in dense aggregates. CLSM did not adequately image the internal filamentous structure and the location of stained cells within dense flocs. However, for typical activated sludge flocs epifluorescence and CLSM proved adequate.

Parallel hybrid modeling methods for a full-scale cokes wastewater treatment plant; Lee DS et al.; Parallel hybrid modeling methods are applied to a full-scale cokes wastewater treatment plant. Within the hybrid model structure, a mechanistic model specifies the basic dynamics of the relevant process and a non-parametric model compensates for the inaccuracy of the mechanistic model. First, a simplified mechanistic model is developed based on Activated Sludge Model No. 1 and the specific process knowledge of the cokes wastewater treatment process. Then, the mechanistic model is combined with five different non-parametric models - feedforward back-propagation neural network, radial basis function network, linear partial least squares (PLS), quadratic PLS and neural network PLS (NNPLS) - in parallel configuration. These models are identified with the same data obtained from the plant operation to predict dynamic Behaviour of the process.

The performance of each parallel hybrid model is compared based on their ease of model building, prediction accuracy and interpretability. For this application, the parallel hybrid model with NNPLS as non-parametric model gives better performance than other parallel hybrid models. In addition, the NNPLS model is used to Analyse the Behaviour of the operation data in the reduced space and allows for fault detection and isolation.

Recovery of an environmental chlamydia strain from activated sludge by co-cultivation with Acanthamoeba sp; Collingro A et al.; Chlamydiae are a unique group of obligate intracellular bacteria comprising important pathogens of vertebrates as well as symbionts of free-living amoebae.

Although there is ample molecular evidence for a huge diversity and wide distribution of chlamydiae in nature, environmental chlamydiae are currently represented by only few isolates. This paper reports the recovery of a novel environmental chlamydia strain from activated sludge by co-cultivation with Acanthamoeba sp. The recovered environmental chlamydia strain UV-7 showed the characteristic morphology of chlamydial developmental stages as revealed by electron microscopy and was identified as a new member of the family Parachlamydiaceae (98.7 % 16S rRNA sequence similarity to Parachlamydia acanthamoebae).

Infection studies suggested that Parachlamydia sp. UV-7 is not confined to amoeba hosts but is also able to invade mammalian cells. These findings outline a new straightforward approach to retrieving environmental chlamydiae from nature without prior, tedious isolation and cultivation of their natural host cells, and lend further support to suggested implications of environmental chlamydiae for public health.

Fenton peroxidation improves the drying performance of waste activated sludge; Dewil R et al.; Advanced sludge treatment processes (AST) reduce the amount of sludge produced and improve the dewaterability, thus probably also affecting the heat transfer properties and the drying characteristics of the sludge. This paper studies the influence of the Fenton peroxidation on the thermal conductivity of the sludge.

Results demonstrate that the Fenton's peroxidation

positively influences the sludge cake consistency and hence enhances the mechanical dewaterability and the drying characteristics of the dewatered sludge. For the two sludges used in this study, i.e. obtained from the wastewater treatment plants (WWTP) of Tienen and Sint-Niklaas - the dry solids content of the mechanically dewatered sludge increased from 22.5% to 40.3% and from 18.7% to 35.2%, respectively. The effective thermal conductivity k(e) of the untreated and the peroxidized sludges is measured and used to determine the heat transfer coefficient h(s).

An average improvement for k(e) of 16.7% (Tienen) and 5.8% (Sint-Niklaas) was observed. Consequently the value of h(s) increased with 15.6% (Tienen) and 5.0% (Sint-Niklaas). This increased heat transfer coefficient in combination with the increased dewaterability has direct implications on the design of sludge dryers. A plate-to-plate calculation of a multiple hearth dryer illustrates that the number of plates required to dry the peroxidized sludge to 90% DS is less than half the number of plates needed to dry untreated sludge. This results in reduced dryer dimensions or a higher capacity for an existing dryer of given dimensions.

Discrepancies in the widely applied GAM42a fluorescence in situ hybridisation probe for Gammaproteobacteria; Siyambalapitiya N et al.; A bacterial culture collection of 104 strains was obtained from an activated sludge wastewater treatment plant to pursue studies into microbial flocculation. Characterisation of the culture collection using a polyphasic approach indicated seven isolates, phylogenetically affiliated with the deep-branching Xanthomonas group of the class Gammaproteobacteria, were unable to hybridise the GAM42a fluorescence in situ hybridisation (FISH) probe for Gammaproteobacteria.

The sequence of the GAM42a probe target region in the 23S rRNA gene of these isolates was determined to have mismatches to GAM42a. Probes perfectly targeting the mismatches (GAM42a_T1038_G1031, and GAM42a_T1038 and GAM42a_A1041_A1040) were synthesised, and used in conjunction with GAM42a in FISH to study the

Gammaproteobacteria community structure in one full-scale activated sludge plant. Several bacteria in the activated sludge biomass bound the modified probes demonstrating their presence and the fact that these Gammaproteobacteria have been overlooked in community structure analyses of activated sludge.

Toxicity of metals and organic chemicals evaluated with bioluminescence assays; Ren S et al.; The development of a bioluminescent sensor organism (Shk1) that was created for assessing wastewater toxicity was reported several years ago. In order to establish a test battery to better characterize wastewater toxicity, additional luminescent sensor organisms were later created. The present study focused on one promising candidate (PM6), a Pseudomonas spp. strain, because of its high level of luminescence compared to that of other newly created organisms. Using a batch toxicity testing protocol, the toxicity of 7 metals and 25 organic compounds was evaluated with the PM6 and Shk1 assays.

Results indicated that the toxicity data of the PM6 and the Shk1 assays were correlated, and no assay appeared to be particularly more sensitive to a group of toxicants than the other assay.

The results of the PM6 and Shk1 assays were further evaluated by comparing with the results of the Vibrio fischeri luminescence inhibition assay and activated sludge inhibition assays. Data suggested that PM6 and Shk1 more closely represented activated sludge organisms than V. fischeri. The suitability of using PM6 and Shk1 for assessing wastewater toxicity on activated sludge, both individually and in a test battery, was discussed.

Aerobic biodegradation of organotin compounds in activated sludge batch reactors; Stasinakis AS et al.; The biodegradation Behaviour of four organotin (OT) compounds, namely tributyltin (TBT), dibutyltin (DBT), monobutyltin (MBT) and triphenyltin (TPhT), was studied in lab-scale activated sludge batch reactors. The activated sludge was spiked with the OT compounds at a level of 100mugl(−1) as Sn. Determination of the OT compounds by GC-FPD after ethylation in the dissolved and particulate phase revealed that 24h after the start of the experiments, almost the total of OT compounds has been removed from the dissolved phase and is associated with the suspended solids. Calculation of mass balance in batch reactors showed that OT compounds biodegradation was performed via a sequential dealkylation process.

Removals due to biodegradation were differentiated according to the parent compound. In experiments with non-acclimatized biomass, a percentage of 27.1, 8.3, 73.8 and 51.3 was still present as TBT, DBT, MBT and TPhT, respectively, at the end of the experiment (18th day). Half-lives (t(1/2)) of 10.2 and 5.1 days were calculated for TBT and DBT, respectively, whereas apparent t(1/2) values could not be determined for MBT and TPhT (t(1/2)>18 days).

The capacity of activated sludge to biodegrade OT compounds in the absence of supplemental substrate indicated that these compounds can be metabolized as single sources of carbon and energy in activated sludge systems. Excluding TBT, the presence of low concentrations of supplemental substrate did not affect the biodegradation potential of activated sludge.

The acclimatization of biomass on OT compounds enhanced significantly biodegradation, resulting in significant decreases of half-lives of OT compounds. As a result in the presence of acclimatized biomass, half-lives of 1.4, 3.6, 9.8 and 5.0 days were calculated for TBT, DBT, MBT and TPhT, respectively.

Genes Involved in Aniline Degradation by Delftia acidovorans Strain 7N and Its Distribution in the Natural Environment; Urata M et al.; Aniline-degraders were isolated from activated sludge and environmental samples and classified into eight phylogenetic groups.

Seven groups were classified into Gram-negative bacteria, such as Acidovorax sp., Acinetobacter sp., Delftia sp., Comamonas sp., and Pseudomonas sp., suggesting the possible dominance of Gram-negative aniline-degraders in the environment. Aniline degradative genes were cloned from D. acidovorans strain 7N, and the nucleotide sequence of the

8,039-bp fragment containing eight open reading frames was determined.

Their deduced amino acid sequences showed homologies to glutamine synthetase (GS)-like protein, glutamine amidotransferase (GA)-like protein, large and small subunits of aniline dioxygenase, reductase, LysR-type regulator, small ferredoxin-like protein, and catechol 2,3-dioxygenase, suggesting a high similarity of this gene cluster to those in P. putida strain UCC22 and Acinetobacter sp. strain YAA. Polymerase chain reaction (PCR) and sequencing analyses of GS-like protein gene segments of other Gram-negative bacteria suggested that Gram-negative bacteria have aniline degradative gene that can be divided into two distinctive groups.

Leucobacter chromiireducens sp. nov, and Leucobacter aridicollis sp. nov., two new species isolated from a chromium contaminated environment; Morais PV et al.; Two strains designated strains L-1T and L-9T were isolated from activated sludge of a treatment plant that receives wastewater from the tannery industry contaminated with chromium.

Phylogenetic analysis showed that the organisms represented two new species of the genus Leucobacter. Strains L-1T and L-9T could be distinguished from the type strain of L. komagatae and from the type strain of"L. albus"by the B-type peptidoglycan composition, fatty acid composition, several phenotypic and physiological characteristics. The major fatty acids of the organisms were iso- and anteiso-branched C15:0 and C17:0, straight-chain C16:0 was also found in relatively high proportions.

The organisms were halotolerant, grew in medium containing 9% NaCl, and all strains, including the type strain of L. komagatae grew in medium containing 5 mM Cr(VI). On the basis of the distinct peptidoglycan composition, 16S ribosomal DNA sequence analysis, percentage of DNA-DNA reassociation values, and phenotypic characteristics we are of the opinion that strain L-1T represents a new species of the genus Leucobacter for which we propose the name Leucobacter chromiireducens and that strain L-9T represents an additional new species of the same genus for which we propose the name Leucobacter aridicollis.

The solids retention time-a suitable design parameter to evaluate the capacity of wastewater treatment plants to remove micropollutants; Clara M et al.; Micropollutants as endocrine disrupting compounds (EDC) or pharmaceuticals are of increased interest in water pollution control. Wastewater treatment plant (WWTP) effluents are relevant point sources for residues of these compounds in the aquatic environment. The solids retention time (SRT) is one important parameter for the design of WWTPs, relating to growth rate of microorganisms and to effluent concentrations.

If a specific substance is degraded in dependency on the SRT, a critical value for the sludge age can be determined. In WWTPs operating SRTs below this critical value, effluent concentrations in the range of influent concentrations or a distribution according to the adsorption equilibrium have to be expected, whereas in WWTPs operating at SRTs higher than the critical value degradation will occur. Critical SRTs were determined for different micropollutants, indicating that the design criteria based on the sludge age allows an estimation of emissions. Different treatment technologies as conventional activated sludge systems and a membrane bioreactor were considered and no significant differences in the treatment efficiency were detected when operated at comparable SRT.

The results of the investigations lead to the conclusion that low effluent concentrations can be achieved in WWTPs operating SRTs higher than 10 days (referred to a temperature of 10 degrees C). This corresponds to the requirements for WWTPs situated in sensitive areas according to the urban wastewater directive of the European Community (91/271/EEC) in moderate climatic zones.

Reduction of bromate to bromide coupled to acetate oxidation by anaerobic mixed microbial cultures; van Ginkel CG et al.; Bromate, a weakly mutagenic oxidizing agent, exists in surface waters. The biodegradation of bromate was investigated by assessing the ability of mixed cultures of micro-organisms for utilization of bromate as electron acceptor and acetate as electron donor. Reduction of bromate was only observed at relatively low concentrations (<3.0mM) in the absence of molecular oxygen. Under these conditions bromate was reduced stoichiometrically to bromide.

Unadapted sludge from an activated sludge treatment plant and a digester reduced bromate without lag period at a constant rate. Using an enrichment culture adapted to bromate, it was demonstrated that bromate was a terminal electron acceptor for anaerobic growth. Approximately 50%of the acetate was utilized for growth with bromate by the enrichment culture. A doubling of 20h was estimated from a logarithmic growth curve. Other electron acceptors, like perchlorate, chlorate and nitrate, were not reduced or at negligible rates by bromate-utilizing microorganisms.

Fate of Ah receptor agonists during biological treatment of an industrial sludge containing explosives and pharmaceutical residues; Gustavsson LK et al.; GOAL, SCOPE AND BACKGROUND: Sweden is meeting prohibition for deposition of organic waste from 2005. Since 1 million tons of sludge is produced every year in Sweden and the capacity for incineration does not fill the demands, other methods of sludge management have to be introduced to a higher degree. Two biological treatment alternatives are anaerobic digestion and composting.

Different oxygen concentrations result in different microbial degradation pathways and, consequently, in a different quality of the digestion or composting residue, It is therefore necessary to study sludge treatment during different oxygen regimes in order to follow both degradation of compounds and change in toxicity.

In this study, an industrial sludge containing explosives and pharmaceutical residues was treated with anaerobic digestion or composting, and the change in toxicity was studied. Nitroaromatic compounds, which are the main ingredients of both pharmaceutical and explosives, are well known to cause cytotoxicity and genotoxicity.

However, little data are available concerning sludge with nitroaromatics and any associated dioxin-like activity.

Therefore, we studied the sludge before and after the treatments in order to detect any changes in levels of Ah receptor (AhR) agonists using two bioassays for dioxin-like compounds. METHODS: An industrial sludge was treated with anaerobic digestion or composting in small reactors in a semi-continuous manner.

The same volume as the feeding volume was taken out daily and stored at –20 degrees C. Sample preparation for the bioassays was done by extraction using organic solvents, followed by clean up with silica gel or sulphuric acid, yielding two fractions. The fractions were dissolved in DMSO and tested in the bioassays. The dioxin-like activity was measured using the DR-CALUX assay with transfected H4IIE rat hepatoma pGudluc cells and an EROD induction assay with RTL-W1 rainbow trout liver cells.

Reduction of excess sludge production by 3,3',4', 5-tetrachlorosalicylanilide in an activated sludge process; Ye FX et al.; The potential of 3,3',4',5-tetrachlorosalicylanilide (TCS) addition to an activated sludge continuous process to reduce excess sludge production by disrupting coupling between anabolism and catabolism was investigated. TCS was chosen as a metabolic uncoupler for continuous test in a lab-scale completely mixed activated sludge process. TCS reduced sludge yield by approximately 30% at a dosage of 40 mg/day. Substrate removal capability was not adversely affected by the presence of TCS, but effluent nitrogen concentration increased during the 60-day continuous operation. Sludge settleability of treated and control samples was qualitatively comparable and not significantly different. Microbial activities in terms of specific oxygen uptake rate were also enhanced, and the microbial population was altered. The results suggest that TCS is an effective chemical uncoupler that reduces sludge yield; process performance was not significantly affected by introduction of the uncoupler.

Stability of the fluorogenic enzyme substrates and pH optima of enzyme activities in different Finnish soils; Niemi RM et al.; Fluorogenic artificial substrates facilitate sensitive enzyme activity measurements for a variety of processes in soil and other environmental samples. It is possible to use in situ pH for measurements on condition that the substrates are chemically stable. We studied the stability of 12 different methyl umbellipherone (MUF) and amino methyl coumarine (AMC) derivatives used as substrates for arylsulphatase, alpha-glucosidase, beta-glucosidase, beta-xylosidase, cellobiosidase, chitinase, phosphomonoesterase (PME), phoshodiesterase (PDE), esterase, lipase and alanine- and leucine aminopeptidases (AP) over the pH range from 4.0 to 8.0 in modified universal buffer (MUB).

Stability of the substrates for lipase (4-MUF-heptanoate) and esterase (4-MUF-acetate) measurements was poor, especially at the higher pH values. Chitinase substrate, 4-MUF-N-acetyl-beta-d-glucosamide, was unstable at high pH values whereas the substrate for PME activity measurement (4-MUF-phosphate) disintegrated at low pH.

The other substrates and MUF and AMC standard solutions were stable over the pH range studied. The optima between pH 4 and 8 of the 11 different enzyme activities were measured in three forest and two agricultural soil samples and in one activated sludge sample. In soil, for alanine and leucine AP the pH optima were usually 7.5 or higher, for arylsulphatase, beta-glucosidase, beta-xylosidase, esterase and PDE between 4 and 5.5, and for cellobiosidase between 4 and 5. alpha-Glucosidase had an optimum below 5.5 but also exhibited high activity at pH 7.

Soil-dependent variation in pH optima were observed for chitinase, esterase, PDE and PME. Enzyme activities were also measured in 0.5 M acetate buffer at pH 5.5. This buffer yielded the highest activities in all soil samples for arylsulphatase, PDE and PME.

Persistent organic pollutants (POPs) in the conventional activated sludge treatment process: fate and mass balance; Katsoyiannis A et al.; The fate and the mass balance of persistent organic pollutants (POPs) during the conventional activated sludge treatment process were investigated in the wastewater treatment plant of the city of Thessaloniki, northern Greece. The POPs of interest were 7 polychlorinated biphenyls and 19 organochlorine pesticides. Target compounds were determined at six different points across the treatment system: the influent, the effluent of the primary sedimentation tank, the effluent of the secondary sedimentation tank, the primary sludge, the activated sludge from the recirculation stream, and the digested/dewatered sludge.

The distribution of POPs between the dissolved and the adsorbed phases of wastewater and sludge was investigated. A good linear relationship between the distribution coefficients, K(d), and the octanol-water partition coefficients, K(ow), of the solutes was observed only in raw wastewater, suggesting that other factors affect the phase distribution of organic compounds in treated wastewater. For all POPs, a significant increase in partitioning with a decreasing solids concentration was observed, revealing an effect from non-settling microparticles remaining in the "dissolved" phase during the separation procedure.

A good linear relationship was also revealed between logK(d) and the dissolved organic carbon (DOC) content of wastewater, suggesting that DOC Favours the advective transport of POPs in the dissolved phase. Almost all POPs showed good mass balance agreements at both the primary and the secondary treatment.

The losses observed for some species could be attributed to biodegradation/biotransformation rather than volatilization. The relative distribution between the treated effluent and the waste sludge streams varied largely among different compounds, with p-p'-DDE being highly accumulated in the waste sludge (98%) but almost 60% of alpha-HCH remaining in the treated effluent.

Sludge as source of energy and revenue; Onyeche TI; Sludge is a residue/product from wastewater treatment plants and contains most of the contaminants released during human activities.

Some stringent environmental regulations on sludge treatment and disposal exist in many countries. This has resulted in increasing interest in sludge treatment methods that encourage sludge reduction and improvement in biogas production during anaerobic digestion processes. This work demonstrates the first exploitation of valuable energy from homogenised sludge at technical scale with mass reduction.

The optimal combination of sludge homogenisation at relatively low pressures using a modified high-pressure homogeniser led to the success of this unique project. Results showed that about 30% more energy could be obtained from thickened and disrupted sludge than from untreated samples. The energy produced was higher than that invested during disruption and digestion processes. About 23% sludge reduction was also observed with no increase in chemical oxygen demand.

This new process can produce extra energy for local electrification and heating the digester while the sludge reduction provides economic benefits. Concentration of sludge causes reduction in investment cost on digester as well as reduction in operational time for sludge dewatering.

Direct generation of electricity from sludges and other liquid wastes; Dentel SK et al.; Recently it has been shown that electrical energy can be harvested from marine sediments, simply by connection of an electrode (anode) in anaerobic marine sediments to an electrode in the aerobic zone above the sediments. We have now shown that similar applications are available in sludge treatment. Using a reactor with graphite foil electrodes in an aerated aerobic and anaerobic sludge zone, electrical current was generated, and enhanced when an additional organic substrate (acetate) was added. Electron microscopy, x-ray diffraction, and PCR examination of the anode surface showed no surface colonization and no increase in Geobacter relative to a control, indicating that microbial use of the anode as an electron acceptor wa indirect through the use of redox mediators. Given the demonstration of electricity generation from sludge, the potential for similar applications, using other organic waste sources, is evaluated.

Physical characteristics of a waste activated sludge: conditioning responses and correlations with a synthetic surrogate; Dursun D et al.; The efficient and economic management of waste activated sludge (WAS) requires a proper understanding of the sludge's material properties. Though there has been much study of WAS, an adequate linkage between its physical and chemical properties has been elusive. In particular, the conditioning and dewatering of WAS are expensive operations, and the addition of polymer to WAS leads to even more complex material Behaviour that is difficult to optimize. This paper reports on an extensive characterization of WAS, both with and without polymer conditioning. We combined the classical "jar test" approach with less conventional rheometric and electrokinetic measurements. In addition to the use of sampled WAS, a synthetic surrogate sludge was formulated, attempting to duplicate properties of WAS and allow more extensive characterization of a reproducible surrogate to WAS. Results with both the synthetic surrogate and WAS indicated that the traditional, electrokinetic, and rheological properties were related. However, the dose optima by rheometry were somewhat higher in both cases.

Minimization of sludge production and stable operational condition of a submerged membrane activated sludge process; Bhatta CP et al.; Modification of conventional activated sludge process by adding microfiltration type hollow fibre membrane module offers various advantages. The membrane module is immersed directly in the bioreactor to separate the treated water from the sludge and the modified process is called submerged membrane activated sludge process (SMASP). Bench scale SMASP units were operated at constant flux to study the possibility of minimizing the excess sludge production and stable operation at different SRT and BOD loading.

The long SRT developed high stabilized MLSS concentration in the bioreactor and decreased the sludge loading rate {kg-BOD/kg-MLSS-d}. In SMASP, very low sludge yield coefficients (0.04-0.09 kg-MLSS/kg-BOD) were obtained at long SRT (500 d) and small BOD loading (< or =0.5 kg/(m3 x d)) as compared to conventional activated sludge process and these coefficients showed considerable minimization in excess activated sludge production. At the same operation conditions, suction pressure was also maintained low (<25 kPa), and then SMASP was operated stably for long time (approximately 200 d). Co-conditioning and dewatering of alum sludge and waste activated sludge; Lai JY et al.; Co-conditioning and dewatering Behaviours of alum sludge and waste activated sludge were investigated. Two different sludges were mixed at various ratios (2:1; 1:1; 1:2; 1:4) for study. Capillary suction time (CST) and specific resistance to filtration (SRF) were utilized to assess sludge dewaterability. Relatively speaking, waste activated sludge, though of higher solid content, was more difficult to be dewatered than alum sludge.

It was found that sludge dewaterability and settlability became better with increasing fraction of alum sludge in the mixed sludge. Dosage required of the cationic polyelectrolyte (KP-201C) for dewatering was reduced as well. It is proposed that alum sludge acts as skeleton builder in the mixed sludge, and renders the mixed sludge more incompressible which is beneficial for sludge dewatering. e, a, g. Implications of the results of the study to the sludge management plan for Taipei City that generates both alum sludge and waste activated sludge at significant amount are also discussed. The current sludge treatment and disposal plan in Metropolitan Taipei could be made more cost-effective.

Ultrasonic sludge treatment for enhanced anaerobic digestion; Hogan F et al.; Ultrasound is the term used to describe sound energy at frequencies above 20 kHz. High-powered ultrasound can be applied to a waste stream via purpose-designed tools in order to induce cavitation. This effect results in the rupture of cellular material and reduction of particle size in the waste stream, making the cells more amenable to downstream processing. sonix is a new technology utilising high-powered, concentrated ultrasound for conditioning sludges prior to further treatment. This paper presents recent results from a number of demonstration and full-scale plants treating thickened waste activated sludge (TWAS) prior to anaerobic digestion, therefore enhancing the process. The present studies have proved that the use of ultrasound to enhance anaerobic digestion can be achieved at full scale and effectively result in the TWAS (typically difficult to digest) behaving, after sonication, as if it were a"primary"sludge. The technology presents benefits in terms of increased biogas production, better solids reduction, improved dewatering characteristics of the digested sludge mixture and relatively short payback periods of two years or less subject to the site conditions and practices applicable at that time.

Three-dimensional dual-morphotype species modeling of activated sludge flocs; Martins AM et al.; An individual-based model, originally developed for a biofilm system, was adapted to simulate three-dimensional formation of activated sludge flocs. The model was extended to two different bacterial morphotypes (floc-forming and filamentous bacteria), allowing spatial development of the floc according to the bacterial morphology, diffusion, reaction, and growth processes.

The model needed also extension with a process for attachment of individual cells. Despite being in an early stage of development, the model is already a tool that enables us to obtain useful information about the microfloc environment. The model indicates that filamentous bacterial morphology and substrate microgradients are important aspects in the formation of bacterial structures.

In mass transport-limited regimes filamentous bacterial structures prevail, whereas in growth-limited regimes irregular shaped flocs with fingerlike structures are dominant. These modeling results suggest that activated sludge flocs and biofilms might be different manifestations of the same phenomena. The model results support the hypothesis that floc-macrogradients can be the most important parameter for development of bulking sludge.

The model suggests that attachment has a very strong effect on floc structure, leading to enhancement of the effect of substrate microgradients.

Physicochemical properties and stability of activated sludge flocs under temperature upshifts from 30 to 45 degrees

C; Morgan-Sagastume F et al.; The impacts of temperature shifts from 30 to 45 degrees C on the structural stability and surface charge of activated sludge flocs were assessed in four sequencing batch reactors (SBRs) treating pulp and paper mill effluent.

The improvement in floc stability was tested by sludge magnesium enrichment in one SBR and by operating another reactor at a high sludge retention time (SRT) of 33 days. Floc stability was characterized by dissociation constants with solutions of CaCl(2), KCl, urea, and ethylenediamine tetraacetate (EDTA). Surface charge was assessed by cationic-anionic titration and metals concentrations were also determined. The temperature shift consistently caused an increase in the negative sludge surface charge from approximately –0.180 to –0.300 meq/g MLSS.

Magnesium enrichment and a high SRT of 33 days promoted less negatively charged sludge, dampened the increase in negative sludge surface charge, and yielded structurally stronger flocs; however, sludge deflocculation still occurred. Manganese and iron appeared to be released by sludge under the temperature shift. It was concluded that the temperature shift deteriorates the flocculating physicochemical properties of the sludge and that better floc stability achieved by magnesium enrichment and a high SRT is not enough to stop deflocculation. Further research is required to clarify the origin of the increase in negative sludge surface charge, the role of metals, and the governing factors in sludge deflocculation under such temperature shifts.

Optimized sequence retrieval from single bands of temperature gradient gel electrophoresis profiles of the amplified 16S rDNA fragments from an activated sludge system; Zhang X et al.; Sequence retrieval from single bands of polymerase chain reaction (PCR)-denaturing gel electrophoresis (DGE) profiles is an important but often difficult step for molecular diversity analysis of complex microbial communities such as activated sludge systems. We Analysed the temperature gradient gel electrophoresis (TGGE) profiles of PCR-amplified 16S rDNA fragments from an activated sludge sample of a coking wastewater treatment plant. Single bands were excised, and a clone library was constructed for each.

Sequence heterogeneity in each single band was found to be significantly overestimated due to single-stranded DNA (ssDNA) contamination formed during the PCR amplification, since only 10-60% of library clones of each single TGGE band had identical migration Behaviour compared with the parent band. Three methods, digestion with mung bean nuclease, optimization of PCR amplification, and purification via denatured polyacrylamide gel electrophoresis (d-PAGE), were compared for their ability to minimize ssDNA contamination, with the last one being the most efficient. After using d-PAGE to minimize ssDNA to a nearly nondetectable level, 70-100% of library clones for each single TGGE band had identical migration compared with the parent band.

Several sequences were found in each of six single bands, and this co-migration could be predicted with the Poland software. The predominant bacteria of the activated sludge were assessed via a combination of sequence retrieval from each single TGGE band and band intensity analysis. Only beta and alpha subclasses of the Proteobacteria were detected, 93.8% and 6.2%, respectively. Our work suggests that prior to constructing a clone library to retrieve the actual sequence diversity of a single DGE band, it is advisable to minimize ssDNA contamination to a nondetectable level.

High rate treatment by aerobic upflow sludge blanket (AUSB) with external oxygenation; Sharma KR et al.; A 3-year study was conducted using an aerobic upflow sludge blanket (AUSB) reactor to achieve high-rate biological treatment through maintenance of a high biomass concentration (7-10 g VSS/L) together with a high oxygen flux. The AUSB reactor was not aerated directly; instead, oxygenation was provided in an external chamber with flow recirculation. The oxygenation was provided at four different pressurizations: 0, 15, 20 and 25 psig. The AUSB reactor was also pressurized to avoid the flotation of biomass. The flow recirculation rate was varied from 400% to 1,500% in order to ensure adequate oxygen supply. It was found that the AUSB system was able to handle a volumetric loading of as high as 10 kg COD/m3-day with a removal efficiency of 92%. Despite a high upflow velocity through AUSB, the effluent suspended solid concentration was mostly below 60 mg/L. The active fraction of biomass in the AUSB sludge was about 3 times higher than that of the regular activated sludge. This was indicated by a very high specific oxygen uptake rate (SOUR), up to 180-250 mg O2/g VSS-hr. The sludge yield in the entire system was only 0.09 to 0.13 g VSS/g COD removed. This was mainly caused by additional auto-oxidation of biosolids in the oxygenation chamber due to flow recirculation.

Effect of chlorination bulking control on water quality and phosphate release/uptake in an anaerobic-oxic activated sludge system; Chang WC et al.; This study evaluates the effect of chlorination bulking control on water quality and phosphate release/uptakc in an anaerobic-oxic activated sludge system. A series of batch experiments with different specific NaOCl mass dose were conducted to determine the sludge settling properties, supernatant water quality and phosphate metabolism Behaviour of filamentous bulking sludge.

The harvested sludge was from a continuous-flow anaerobic-oxic (A/O) activated sludge pilot-plant, i.e., enhanced biological phosphorus removal (EBPR) system, operated with 15 days of sludge retention time. The filamentous bacteria in the A/O pilot plant were identified to be Thiothrix according to Eikelboom's classification techniques, which was in accordance with the high influent sulfate concentration of this study (50 mg/L sulfate). Increasing NaOCI concentration, as revealed by experimental results, obviously decreased the sludge settling properties (SVI values and zone settling velocities) and meanwhile significantly reduced supernatant water quality (COD, SS, TP) mainly due to higher suspended solids caused by floc disruption. Moreover, the nine-hour batch experiments indicated that high NaOCI dosage (40 mg/gMLSS) completely deteriorated phosphate metabolism of EBPR sludge. Such a high dosage of chlorination further confirmed overdosing through disappearance of intracellular PHB and death of protozoa by microscopic investigation.

Still, phosphate release/uptake Behaviour of EBPR sludge properly functions at low NaOCl dosage (5 mg/g MLSS). Besides, phosphate metabolism worsens rapidly before the SVI value reaches its lowest level. These findings imply that determining NaOCI requirement with merely SVI values can readily result in chlorination overdosing. Proper NaOCI dosage requires a delicately balanced consideration between sludge settling improvement, water quality demand and phosphate metabolism. Batch test of phosphate release/uptake is apparently a prerequisite to conclude an appropriate NaOCl dosage for bulking control.

Microbial degradation of estrogens using activated sludge and night soil-composting microorganisms; Shi JH et al.; In order to investigate the potential for microbial degradation of estrogens, and the products formed, activated sludge collected from Korea (ASK) and night soil-composting microorganisms (NSCM) were used to degrade estrogens. Results showed that both ASK and NSCM degraded almost 100% of the natural estrogens estrone (E1), 17beta-estradiol (E2), and estriol (E3) from initial concentrations of 20-25 mg/L, while synthetic estrogen, ethynylestradiol (EE2), was not degraded. Analysis of degradation products of E2 by using HPLC-ECD and a consecutive first-order reaction calculation confirmed that E2 was sequentially degraded to E1, which was further degraded to other unknown compounds by ASK and NSCM.

We then used the yeast two-hybrid assay to show that the unknown degradation products did not appear to possess estrogenic activity when E1, E2 or E3 were degraded to below the detection limit after 14 days of incubation, indicating that ASK and NSCM not only degrade natural estrogens, but also remove their estrogenic activities.

Effects of phosphorus limitation and temperature on PHA production in activated sludge; Chinwetkitvanich S et al.; The study was designed to investigate the effects of temperature and phosphorus limitation on polyhydroxyalkanoate (PHA) production and storage by activated sludge biomass. The two-stage operation approach, i.e. a growth phase followed by a nutrient limitation phase, was applied to induce PHA accumulation.

The pre-selected temperatures of 10, 20 and 30 degrees C were investigated under phosphorus limitation conditions using three four-litre fully aerobic SBR systems operated at an SRT of 10 days with cycle time and HRT of 6 and 10 hours. PHA production was greater in the 10 degrees C system than in the 20 degrees C and 30 degrees C systems but there was little difference between the two higher temperatures. The maximum PHA fractions of the sludge were 52, 45 and 47%TSS for the three temperatures from low to high, and the maximum PHA concentrations in the mixed liquors were 1,491, 1,294 and 1,260 mg/l, respectively.

However, it was observed that very low values of PHA yield per unit COD consumed were obtained, i.e., 0.05, 0.03 and 0.04 mgPHA/mgCODu, for the 10, 20 and 30 degrees C reactors, respectively. This was because all three systems required several days to reach maximum PHA accumulation in their mixed liquor biomasses. It is probable the bacteria still had some stored poly-P in their cells upon initiation of the phosphorus limited influent, and PHA accumulation was delayed until the stored phosphorus was depleted. Also, PHA productivity was reduced by the large amounts of biomass lost from the systems because of sludge bulking.

Sludge thickening performance of mesh filtration process; Park MS et al.; Small-scale wastewater treatment facilities play an important role in improving the aquatic environment in many countries. Although sludge treatment is essential for overall wastewater treatment, it is difficult for small-scale facilities to use mechanical equipment or other facilities. As the first step of the sludge treatment, it is important to develop a convenient sludge thickening process for small-scale facilities. In this work, we examined the sludge thickening performance of a mesh filtration system: the mesh opening sizes of 100-500 microm, and the sludge (3,000-9,000 mg-SS/L) was obtained from a domestic wastewater treatment facility. The filtration was carried out only under the hydraulic pressure between the water level and the effluent port connected to the mesh filter module. The sludge reduction rates were in the range of 85-95% for 6-7 h; the initial filtration rate was very high, but the rate decreased with a decrease in hydraulic pressure due to the reduction of the water level in the vessel. In addition, the effluents (passed through the mesh) contained very low SS and could be directly discharged into the environment.

Biological treatability of raw and ozonated penicillin formulation effluent; Cokgor EU et al.; In the present study, oxidative pre-treatment of pharmaceutical wastewater originating from the formulation of the penicillin Sultamycillin Tosylate Diydrate via ozonation at varying pH and ozone feed rates was investigated. Biological treatability studies were performed with a synthetic wastewater alone and supplemented with raw and ozonated penicillin formulation effluents.

The highest COD (34%) and TOC (24%) removal efficiencies were obtained at pH 11.0, whereas the BOD5 value increased from 16 mg l(–1) to 128 mg l(–1) after 40 min of ozonation, corresponding to an applied ozone dose of 1670 mg l(-1) and 33% relative ozone absorption. The studies showed that no degradation of raw penicillin fraction (30% of total COD) occurred, and degradation of the synthetic wastewater being completely treatable without penicillin addition, was inhibited by 7%. Upon 40 min ozonation, the synthetic wastewater could be completely oxidized and at the same time 35% of ozonated penicillin wastewater removal was obtained.

Respirometric studies were conducted in parallel and produced results indicating a 22% decrease in the total oxygen consumption rate established for raw penicillin formulation effluent compared to the results obtained from the aerobic batch reactor. No inhibition of the synthetic fraction was observed for the 40 min-ozonated penicillin formulation effluent, biodegradability of the 60 min-ozonated penicillin effluent decreased possibly due to recalcitrant oxidation product accumulation. The modeling study provided experimental support and information on inhibition kinetics in activated sludge model no. 3 (ASM3) by means of respirometric tests for the first time.

Reduction of coagulant amount added to activated sludge for phosphorus removal; Nakajima J et al.; Adding coagulant to the activated sludge process is effective in maintaining the stability of phosphorus removal. However, the precise mechanisms of the reaction and Behaviour of coagulants and phosphorus are not well known. By introducing a new phosphorus removal model (PRM), the Behaviour of coagulant and phosphorus in the process could be described. The experimental data of the effluent phosphorus concentration and Fe content in the activated sludge agreed with the values calculated by PRM. The amount of coagulant addition to the activated sludge process for phosphorus removal is reduced with the enhanced biological phosphorus removal process. It is suggested that the amount of reduction is determined by using PRM.

Monitoring activated sludge settling properties using image analysis; Jenne R et al.; The goal of this study is to develop a monitoring system for activated sludge properties, as this is an essential tool in the battle against filamentous bulking. A fully automatic image analysis procedure for recognising and characterising flocs and filaments in activated sludge images has been optimised and subsequently used to monitor activated sludge properties in a lab-scale installation. The results of two experiments indicate that the image information correlates well with the Sludge Volume Index. It is shown that, at the onset of filamentous bulking, there is an increase in total filament length on the one hand, and a change in floc shape on the other hand.

Performances of three R-AN-D-N wastewater treatment plants in the Czech Republic; Bejvl Z et al.; The evolution of Czech standards requires higher efficiency of nutrient removal from municipal wastewaters. At the beginning of the last decade of 20th century, a new activated sludge configuration called R-AN-D-N process has been described, successfully tested and now largely used at several wastewater treatment plants (WWTP) in the Czech republic.

The main feature of the R-AN-D-N process is the introduction of a regeneration zone in sludge recycle, which enables to increase sludge age in the system without any substantial increase in WWTP volume. Performances of three Czech large WWTP with R-AN-D-N configuration have been monitored and compared within a period of one and a half years.

The results confirmed excellent nutrient removal efficiency for wastewaters with different proportion between sewage and industrial effluents. Two of three monitored WWTP received wastewaters from breweries (Budweiser and Pilsner Urquell). The settleability of activated sludge from all three WWTP was correct, with SVI values usually ranging from 50 to 150 ml/g. Monitoring of sludge composition indicated proliferation of several filamentous bacteria, particularly types 0581, 0092 and M. parvicella. No severe bulking events were observed. Finally, the operational costs expressed in CZK (Czech crown: 1 CZK = {see symbol in text}0.0322) per cubic metre of treated sewage or per capita amounts respectively from 2.24 to 6.52, and from 285 to 342.

North American performance experience with anoxic and anaerobic selectors for activated sludge bulking control; Parker D et al.; Obtaining reliably high compaction characteristics and low SVI values has enormous economic consequences on activated sludge plant design and operation. Now that use of selectors has become more common for achieving low SVI values, the assessment of the full-scale performance characteristics of activated sludge plants incorporating selectors has high value for the design and operation of future plants. Treatment plant effluent quality is seldom Analysed solely on an average basis and the impact peak events have on effluent quality must be considered in design and operation. Descriptive statistics should also be used to assess selector performance, so that the influence of infrequently occurring values on plant design and operation can be assessed.

The performance of 21 activated sludge plants incorporating selectors was evaluated. All the selector plants for which post installation data were available showed that the operating SVI values were significantly improved. One plant studied in detail showed that final effluent quality was significantly better after the selector installation. As a group, activated sludge plants with anaerobic selectors outperform those with anoxic selectors. Dissolved oxygen control is just as important in activated sludge plants with selectors as in conventional activated sludge plants.

Influence of sludge properties and hydraulic loading on the performance of secondary settling tanks--full-scale operational results; Vestner RJ et al.; Full-scale investigations at a WWTP with a two-stage secondary settling tank process revealed relationships between significant operating parameters and performance in terms of effluent suspended solids concentration. Besides common parameters (e.g. surface overflow rate and sludge volume loading rate) feed SS concentration and flocculation time must be considered. Concentration of the return activated sludge may help to estimate the performance of existing secondary settling tanks.

Modeling of a large-scale wastewater treatment plant for efficient operation; Gokcay CF et al.; Environmental legislations in the Western world impose stringent effluent quality standards for ultimate protection of the environment. This is also observed in Turkey. The current paper presents efforts made to simulate an existing 0.77 million m3/day conventional activated sludge plant located at Ankara, AWTP. The ASM1 model was used for simulation in this study. The model contains numerous stoichiometric and kinetic parameters, some of which need to be determined on case by case bases. The easily degradable COD (S(S)) was determined by two methods, physical-chemical and respirometric methods, namely. The latter method was deemed unreliable and rejected in the further study. Dynamic simulation with SSSP Programmeme predicted effluent COD and MLSS values successfully while overestimating OUR. A complete fit could only be obtained by introducing a dimensionless correction factor (etaO2 = 0.58) to the oxygen term in ASM1.

A control strategy for reducing aeration costs during low loading periods; Sahlmann C et al.; The efficiency of the aeration system in a full-scale activated sludge basin with 3 separately controlled aeration zones was improved for the low loading period in summer. The air flow rate to each aeration zone is currently regulated to hold a preset dissolved oxygen concentration (DO). Four different DO setpoint combinations were tested, each one for a one week period, using dynamic off-gas testing to measure the standardised oxygen transfer efficiency (alphaSOTE). As the DO setpoints were lowered, the total air flow rate to the basin decreased initially.

A low DO in the first zones slowed biomass activity and pushed the load towards the end of the aeration basin. The relationship between alphaSOTE and the specific diffuser flow rate qD is different for each zone. In Zone 1 there was a strong decrease in alphaSOTE as qD increased, while Zones 2 and 3 were fairly independent of qD, Zone 2 at a higher level than Zone 3. Aeration costs were reduced by 15% for the most efficient combination. To achieve even more savings, a control strategy adjusting oxygen transfer rates over the aeration basin to the necessary oxygen transfer rates is suggested. It is based on changing the DO setpoints to reach the lowest total air flow rate while meeting the effluent requirements.

Nutrient removal process development and full scale implementation at the 4 million p.e. main treatment plant of Vienna, Austria; Kroiss H et al.; The Main Treatment Plant of Vienna is in extension for 4 million p.e. and very stringent nutrient removal requirements. The existing high rate BOD removal activated sludge plant (in operation since 1980) is extended by a second stage activated sludge plant and a newly developed flow scheme for nitrogen removal optimisation adaptable to the temperature variations over the year. For this plant pilot investigations have been performed for the development of a specific mathematical model (ASMV) and a specific aeration control strategy.

The civil work of the extension is already finished and the installation of the equipment has started. Operation should start in 2004. The whole project will cost about {see symbol in text}264 million of which about one half is for civil work. The effluent standards correspond to the requirements for sensitive areas in EU Directive for Municipal Waste Water. The raw primary and excess sludge are incinerated after thickening and dewatering. This paper tries to condense the already existing literature with the construction progress and the cost situation.

Isolation and characterization of sulphur-oxidizing Thiomonas sp. and its potential application in biological deodorization; Chen XG et al.; AIMS: To isolate and characterize a sulphur-oxidizing bacterial strain from activated sludge and to evaluate its potential application in biological deodorization. METHODS AND RESULTS: A dominant sulphur-oxidizing bacterial strain, designated as strain SS, was isolated from an enrichment culture using thiosulphate as a sole energy source and CO2 as a sole carbon source.

The cells of this organism were aerobic, rod-shaped, Gram-negative and motile. Strain SS could grow autotrophically, heterotrophically as well as mixotrophically. Autotrophic growth was observed at pH values ranging from 2.3 to 9.0. Phylogenetic analyses revealed that strain SS belonged to Group 1 of the genus Thiomonas, closely related to Thiomonas perometabolis and Thiomonas intermedia. The thiosulphate oxidation rates of strain SS at different pH values were evaluated in terms of oxygen uptake using a Micro-Oxymax respirometer. The results showed that the maximum oxidation rate of 5.65 mg l(−1) h(−1) occurred at 56 h of growth and pH 6.0. Continuous H2S removal study demonstrated that strain SS could remove more than 99% of H2S when the inlet concentration was below 58.6 ppm.

Further increase of the inlet concentration to 118 ppm gave rise to a decline in the removal efficiency to ca 90%.Conclusions: The strong acidification of the culture medium during the later period could result in the deterioration of the growth activity and the metabolism activity of strain SS. In practical application, the problems caused by the end-product inhibition and the acidification can be alleviated by periodical replacement of culture medium with fresh medium. d, b. Given the physiological flexibility and the ability to remove H2S rapidly and efficiently, strain SS could be a good'deodorizing' candidate. Significance AND

IMPACT OF THE STUDY: This is the first time that Thiomonas species has been reported for biological deodorization application.

Start up of deammonification process in one single SBR system; Li X et al.; A process for autotrophic nitrogen removal named aerobic/anoxic deammonification wherein NH4+ is oxidized by nearly 50% to NO2- and subsequently the ammonia is converted together with the nitrite to molecular nitrogen (N2 gas), has come to full-scale application within the last few years. In this research, sludge from a biological rotation disk located at a landfill leachate plant at Mechernich, Germany, which is capable of performing the deammonification process, was used as seed sludge for acclimating deammonification activities in Labouratory scale batch-reactors.

In parallel, the same tests were performed with normal activated sludge. Research results indicated that deammonification activities could be obtained from the seeded reactor and also, with limited performance, from normal activated sludge in a single SBR system after several months acclimation. It was also seen that oxygen is an important factor that influences the deammonification from both the acclimatization process and process running. Further results were approved that report an impact of nitrite as a process intermediate on the closely related process of anaerobic ammonia oxidation ("Anammox"). However, limiting concentrations on a bacteria population performing deammonification were found to be different to those reported for a pure Anammox-culture. Also the influence of another intermediate, hydrazine, was tested for speeding up the acclimating process by inducing the deammonification activities and recovering the activities of deammonification from nitrite inhibition.

Results of a technology demonstration project to compare rapid aquatic toxicity screening tests in the analysis of industrial effluents; Daniel M et al.; The results of a 'BioWise' demonstration project to assess the comparative sensitivity and practicality of seven new assays for the direct assessment of ecotoxicity in industrial effluents are presented. In addition the aim of the project was to validate the results of the new assays against benchmark data generated from non-proprietary, rapid, microplate screening assays using the regulatory species; freshwater crustacean Daphnia magna and green algae Selenastrum capricornutum, chosen in view of their environmental relevance.

The new commercial test assays were: Daphnia magna, Selenastrum capricornutum and Thamnocephalus platyurus Toxkits supplied by Vickers Labouratories Ltd, containing dormant, immobilised life stages of the test species; GreenScreen EM, a yeast based assay for genotoxicity and general acute toxicity supplied by Gentronix Ltd; and CellSense a mediated, amperometric whole cell biosensor based on immobilised activated sludge and E. coli. 38 effluent samples supplied by members of SOCSA (Specialised Organic Chemicals Sector Association) were examined over a period of 13 months, in the project co-ordinated by the AstraZeneca Brixham Environmental Labouratory, and part funded by BioWise via the UK Government Department of Trade and Industry.

Identification and occurrence of tetrad-forming Alphaproteobacteria in anaerobic-aerobic activated sludge processes; Wong MT et al.; In an acetate-fed anaerobic-aerobic membrane bioreactor, a deteriorated enhanced biological phosphorus removal (EBPR) community was developed (as determined based on the chemical profiles of organic substrate, soluble phosphate, and intracellular carbohydrate and polyhydroxyalkanote (PHA) concentrations).

Microscopic observations revealed the dominance of tetrad-forming organisms (TFOs), of which the majority stained positively for PHA under anaerobic conditions. Fluorescence in situ hybridization (FISH) confirmed that the Alphaproteobacteria (85.0+/–7.0% of total cells) were the most dominant group. A 16S rRNA gene clone library specific for the Alphaproteobacteria indicated that most 16S rRNA gene clones (61% of total clones) were closely affiliated with 'Defluvicoccus vanus', forming a cluster within subgroup 1 of the Alphaproteobacteria. Combined PHA staining and FISH with specific probes designed for the members of the 'Defluvicoccus' cluster suggested diversity within this TFO cluster, and that these TFOs were newly identified glycogen-accumulating organisms in EBPR systems. However, these 'Defluvicoccus'-related TFOs were only seen in low abundance in 12 different EBPR and non-EBPR systems, suggesting that they were not the key populations responsible for the deterioration of full-scale EBPR processes.

Conceptualizing "suicidal genetically engineered microorganisms" for bioremediation applications; Pandey G et al.; Use of genetically modified microorganisms (GEMs) for pollution abatement has been limited because of risks associated with their release in the environment. Recent developments in the area of recombinant DNA technologies have paved the way for conceptualizing "suicidal genetically engineered microorganisms" (S-GEMS) to minimize such anticipated hazards and to achieve efficient and safer bioremediation of contaminated sites. Our strategy of designing a novel S-GEM is based on the knowledge of killer-anti-killer gene(s) that would be susceptible to Programmemed cell death after detoxification of any given contaminated site(s).

Ochrobactrum tritici strain 5bvl1 — characterization of a Cr(VI)-resistant and Cr(VI)-reducing strain; Branco R et al.; Bacterial strain 5bvl1, isolated from a chromium-contaminated wastewater treatment plant and identified as Ochrobactrum tritici, was resistant to a broad range of antibiotics, to Cr(VI), Ni(II), Co(II), Cd(II), and Zn(II), and was able to grow in the presence of 5% NaCl and within the pH range 4–10. Characterization showed that strain 5bvl1 could be considered a halotolerant and alkalitolerant microorganism resistant to high concentrations of Cr(VI). This strain was able to grow aerobically in up to 10 mmol.L–1 Cr(VI). Cr(VI) resistance was independent of sulphate concentration. Under aerobic conditions strain 5bvl1 was also able to reduce high Cr(VI) concentrations (up to 1.7 mmol.L–1). Increasing concentrations of Cr(VI) in the medium lowered the growth rate of strain 5bv11 but the reduction in growth rate could not be directly correlated with the amount of Cr(VI) reduced. Unlike the type strain, which was only able to reduce Cr(VI), strain 5bvl1 was resistant to Cr(VI) and able to reduce it. Moreover, in strain 5bvl1, the rate and extent of Cr(VI)-reduction were higher than in the other strains of the genus Ochrobactrum. Ochrobactrum strain 5bvl1 resists high Cr(VI) concentrations and has a high Cr(VI)-reducing ability, making it a valuable tool in bioremediation.

Bioremediation of a weathered and a recently oil-contaminated soils from Brazil: a comparison study; Trindade PV et al.; The facility with which hydrocarbons can be removed from soils varies inversely with aging of soil samples as a result of weathering. Weathering refers to the result of biological, chemical and physical processes that can affect the type of hydrocarbons that remain in a soil.

These processes enhance the sorption of hydrophobic organic contaminants (HOCs) to the soil matrix, decreasing the rate and extent of biodegradation. Additionally, pollutant compounds in high concentrations can more easily affect the microbial population of a recently contaminated soil than in a weathered one, leading to inhibition of the biodegradation process.

The present work aimed at comparing the biodegradation efficiencies obtained in a recently oil-contaminated soil (spiked one) from Brazil and an weathered one, contaminated for four years, after the application of bioaugmentation and biostimulation techniques. Both soils were contaminated with 5.4% of total petroleum hydrocarbons (TPHs) and the highest biodegradation efficiency (7.4%) was reached for the weathered contaminated soil.

It could be concluded that the low biodegradation efficiencies reached for all conditions tested reflect the treatment difficulty of a weathered soil contaminated with a high crude oil concentration.

Moreover, both soils (weathered and recently contaminated) submitted to bioaugmentation and biostimulation techniques presented biodegradation efficiencies approximately twice as higher as the ones without the aforementioned treatment (natural attenuation).

Biodegradation of phenanthrene by the indigenous microbial biomass in a zinc amended soil; Wong KW et al.; Abstract k.w. wong, b.a. toh, y.p. ting and j.p. obbard. 2004.Aims: To study the effect of zinc on the biodegradation of phenanthrene by the microbial biomass in soil.

Methods and Results: Uncontaminated soil was amended with zinc and phenanthrene as single or co-contaminants, and microbial metabolic activity was measured using an intracellular dehydrogenase enzyme bioassay over 37 days. Contaminants were amended at optimum, action and double the action level specified in 'The New Dutch List' (Ministry of Housing, Spatial Planning and Environment, the Netherlands, 2000). Microbial activity in soils with zinc or phenanthrene alone indicated the presence of tolerant, albeit inhibited soil micro-organisms. A zinc concentration at the optimum level of 140 mg kg(–1) in the co-contaminated soil (phenanthrene at 40 mg kg(–1)) resulted in marginal stimulation of the rate of phenanthrene biodegradation.

However, Zn(2+) concentrations at the action and double the action level of zinc (720 and 1440 mg kg(–1)) inhibited phenanthrene degradation. Conclusions: Biodegradation of phenanthrene in soils co-contaminated with zinc at concentrations above the action value is impeded. Significance and Impact of the Study: Bioremediation efforts to remove polycyclic aromatic hydrocarbon in zinc co-contaminated soils are likely to be constrained.

Stable isotope fractionation analysis as a tool to monitor biodegradation in contaminated acquifers; Meckenstock RU et al.; The assessment of biodegradation in contaminated aquifers has become an issue of increasing importance in the recent years. To some extent, this can be related to the acceptance of intrinsic bioremediation or monitored natural attenuation as a means to manage contaminated sites. Among the few existing methods to detect biodegradation in the subsurface, stable isotope fractionation analysis (SIFA) is one of the most promising approaches which is pronounced by the drastically increasing number of applications. This review covers the recent Labouratory and field studies assessing biodegradation of contaminants via stable isotope analysis.

Stable isotope enrichment factors have been found that vary from no fractionation for dioxygenase reactions converting aromatic hydrocarbons over moderate fractionation by monooxygenase reactions (epsilon=–3 per thousand) and some anaerobic studies on microbial degradation of aromatic hydrocarbons (epsilon=–1.7 per thousand) to larger fractionations by anaerobic dehalogenation reactions of chlorinated solvents (epsilon=between –5 per thousand and -30 per thousand). The different isotope enrichment factors can be related to the respective biochemical reactions. Based on that knowledge, we discuss under what circumstances SIFA can be used for a qualitative or even a quantitative assessment of biodegradation in the environment.

In a steadily increasing number of cases, it was possible to explain biodegradation processes in the field based on isotope enrichment factors obtained from Labouratory experiments with pure cultures and measured isotope values from the field. The review will focus on the aerobic and anaerobic degradation of aromatic hydrocarbons and chlorinated solvents as the major contaminants of groundwater. Advances in the instrumental development for stable isotope analysis are only mentioned if it is important for the understanding of the application.

Biotechnology applied to cultural heritage: biorestoration of frescoes using viable bacterial cells and enzymes; Ranalli G et al.; Abstract g. ranalli, g. alfano, c. belli, g. lustrato, m.p. colombini, i. bonaduce, e. zanardini, p. abbruscato, f. cappitelli and c. sorlini. 2004.Aims: To set up and employ, for the biorestoration of cultural heritage (altered frescoes), an advanced and innovative biotechnology method based on the sequential use of whole viable bacterial cells and specific enzymes. Methods and Results: The bioremediation intervention consisted of the direct application onto an artwork surface of whole bacterial cells of the Pseudomonas stutzeri A29 strain (bioaugmentation), followed by, in a final step, a purified Protease enzyme. The bioremediation was performed on a Spinello Aretino fresco that had become altered by the animal glue residues of past restoration. For the reader's interest the fresco is the 14th century Conversione di S. Efisio e battaglia (Conversion of S. Efisio and battle), size 3.5 × 7.8 m at the Pisa Camposanto Monumentale, Italy.

An assessment was made of the final costs of the biological tests (whole bacterial cells, enzymes) so as to compare them with other intervention techniques. Conclusions: A successful innovative biological approach to recover valuable frescoes was set up, and the best conditions for treatment efficiency were identified.

Furthermore the cost of the biological cleaning using viable bacterial cells and enzymes (P. stutzeri, Protease, Collagenase, 1: 3: 10, ratio respectively) was much lower than that of other conventional methods, making this biotechnology not only very interesting but also very competitive. Significance and Impact of the Study: New biotechnologies with an innovative, soft approach to the 'biocleaning' and 'biorestoration' of cultural heritage are in constant demand, and our results are clear evidence that such an approach has been achieved; the technique could be of significant importance towards developing other goals.

MetaRouter: bioinformatics for bioremediation; Pazos F et al.; Bioremediation, the exploitation of biological catalysts (mostly microorganisms) for removing pollutants from the environment, requires the integration of huge amounts of data from different sources. We have developed MetaRouter, a system for maintaining heterogeneous information related to bioremediation in a framework that allows its query, administration and mining (application of methods for extracting new knowledge).

MetaRouter is an application intended for Labouratories working in biodegradation and bioremediation, which need to maintain and consult public and private data, linked internally and with external databases, and to extract new information from it. Among the data-mining features is a Programmeme included for locating biodegradative pathways for chemical compounds according to a given set of constraints and requirements.

The integration of biodegradation information with the corresponding protein and genome data provides a suitable framework for studying the global properties of the bioremediation network. The system can be accessed and administrated through a web interface. Physiological and molecular genetic analyses of vinyl chloride and ethene biodegradation in Nocardioides sp. strain JS614; Mattes TE et al.; Nocardioides sp. strain JS614 utilizes vinyl chloride and ethene as carbon and energy sources. JS614 could be influential in natural attenuation and biogeochemical ethene cycling, and useful for bioremediation, biocatalysis and metabolic engineering, but a fundamental understanding of the physiological and genetic basis of vinyl chloride and ethene assimilation in strain JS614 is required. Alkene monooxygenase (AkMO) activity was demonstrated in whole-cell assays and epoxyalkane:coenzyme M transferase (EaCoMT) activity was detected in JS614 cell-free extracts.

Pulsed-field gel electrophoresis revealed a 290-kb plasmid (pNoc614) in JS614. Curing experiments and PCR indicated that pNoc614 encodes vinyl chloride/ethene-degradation genes. JS614 vinyl chloride/ethene catabolic genes and flanking DNA (34.8 kb) were retrieved from a fosmid clone. AkMO and EaCoMT genes were found in a putative operon that included CoA transferase, acyl-CoA synthetase, dehydrogenase, and reductase genes.

Adjacent to this gene cluster was a divergently transcribed gene cluster that encoded possible coenzyme M biosynthesis enzymes. a, g, i. Reverse transcription-PCR demonstrated the vinyl chloride- and ethene-inducible nature of several genes. Genes encoding possible plasmid conjugation, integration, and partitioning functions were also discovered on the fosmid clone.

A PCR method for the detection and differentiation of Lentinus edodes and Trametes versiColour in defined-mixed cultures used for wastewater treatment; Garcia-Mena J et al.. A PCR-based method for the quantitative detection of Lentinus edodes and Trametes versiColour, two ligninolytic fungi applied for wastewater treatment and bioremediation, was developed. Genomic DNA was used to optimize a PCR method targeting the conserved copper-binding sequence of laccase genes.

The method allowed the quantitative detection and differentiation of these fungi in single and defined-mixed cultures after fractionation of the PCR products by electrophoresis in agarose gels. Amplified products of about 150 bp for L. edodes, and about 200 bp for T. versiColour were purified and cloned. The PCR method showed a linear detection response in the 1.0 mug-1 ng range.

The same method was tested with genomic DNA from a third fungus (Phanerochaete chrysosporium), yielding a fragment of about 400 bp. Southern-blot and DNA sequence analysis indicated that a specific PCR product was amplified from each genome, and that these corresponded to sequences of laccase genes.

This PCR protocol permits the detection and differentiation of three ligninolytic fungi by amplifying DNA fragments of different sizes using a single pair of primers, without further enzymatic restriction of the PCR products. This method has potential use in the monitoring, evaluation, and improvement of fungal cultures used in wastewater treatment processes.

The enzymes of oxalate metabolism: unexpected structures and mechanisms; Svedruzic D et al.; Oxalate degrading enzymes have a number of potential applications, including medical diagnosis and treatments for hyperoxaluria and other oxalate-related diseases, the production of transgenic plants for human consumption, and bioremediation of the environment.

This review seeks to provide a brief overview of current knowledge regarding the major classes of enzymes and related proteins that are employed in plants, fungi, and bacteria to convert oxalate into CO(2) and/or formate. Not only do these enzymes employ intriguing chemical strategies for cleaving the chemically unreactive C-C bond in oxalate, but they also offer the prospect of providing new insights into the molecular processes that underpin the evolution of biological catalysts.

Resistance of solid-phase U(VI) to microbial reduction during in situ bioremediation of uranium-contaminated groundwater; Ortiz-Bernad I et al.; Speciation of solid-phase uranium in uranium-contaminated subsurface sediments undergoing uranium bioremediation demonstrated that although microbial reduction of soluble U(VI) readily immobilized uranium as U(IV), a substantial portion of the U(VI) in the aquifer was strongly associated with the sediments and was not microbially reducible. These results have important implications for in situ uranium bioremediation strategies.

Appl Environ Microbiol, 2004 Dec, 70(12), 7251 - 9 In situ expression of nifD in Geobacteraceae in subsurface sediments; Holmes DE et al.; In order to determine whether the metabolic state of Geobacteraceae involved in bioremediation of subsurface sediments might be inferred from levels of mRNA for key genes, in situ expression of nifD, a highly conserved gene involved in nitrogen fixation, was investigated.

When Geobacter sulfurreducens was grown without a source of fixed nitrogen in chemostats with acetate provided as the limiting electron donor and Fe(III) as the electron acceptor, levels of nifD transcripts were 4 to 5 orders of magnitude higher than in chemostat cultures provided with ammonium. In contrast, the number of transcripts of recA and the 16S rRNA gene were slightly lower in the absence of ammonium.

The addition of acetate to organic- and nitrogen-poor subsurface sediments stimulated the growth of Geobacteraceae and Fe(III) reduction, as well as the expression of nifD in Geobacteraceae. Levels of nifD transcripts in Geobacteraceae decreased more than 100-fold within 2 days after the addition of 100 microM ammonium, while levels of recA and total bacterial 16S rRNA in Geobacteraceae remained relatively constant. Ammonium amendments had no effect on rates of Fe(III) reduction in acetate-amended sediments or toluene degradation in petroleum-contaminated sediments, suggesting that other factors, such as the rate that Geobacteraceae could access Fe(III) oxides, limited Fe(III) reduction. These results demonstrate that it is possible to monitor one aspect of the in situ metabolic state of Geobacteraceae species in subsurface sediments via analysis of mRNA levels, which is the first step toward a more global analysis of in situ gene expression related to nutrient status and stress response during bioremediation by Geobacteraceae.

Enzymatic bioremediation: from enzyme discovery to applications; CSIRO Entomology, Canberra, Australian Capital Territory, Australia1. Enzymatic bioremediation is potentially a rapid method of removing environmental pesticide residues. Applications include the treatment of residues resulting from agricultural production and processing industries, such as the treatment of irrigation waters, surface-contaminated fruit and vegetables and spent dip liquors. 2.

A specific application for some organophosphate-degrading enzymes involves detoxification of nerve agent stockpiles. Effective and affordable remediation requires highly specialized enzymes, so protein engineering techniques are being used to improve properties of various source enzymes to enhance catalytic rates, stability and substrate range. 3.

Trials with an optimized organophosphate-degrading enzyme have shown the feasibility of such technology in various applications. 4. The enzymes developed for environmental remediation for specific pesticide classes also have applications as antidotes for high-dose pesticide poisonings and as prophylaxis for people at risk of high pesticide doses.

Modeling natural attenuation of chlorinated ethenes under spatially varying redox conditions; Widdowson MA; A three-dimensional model for the transport and reductive dechlorination of chlorinated ethenes in ground-water systems with variable redox conditions is demonstrated and applied to a pilot test for accelerated natural attenuation of trichloroethene (TCE). The rate and extent of biotransformation of TCE and chlorinated progeny is controlled by the dominant terminal electron accepting process (TEAP) that is simulated over space and time.

The solute transport code, Sequential Electron Acceptor Model, 3D-transport, (SEAM3D) which simulates aerobic and sequential anaerobic biodegradation of organic carbon, is modified to implement the equations.

Results of a generic model for TCE transport in groundwater systems with different redox conditions demonstrate that the degree of chlorinated ethene attenuation is influenced by background concentrations of aqueous- and solid-phase electron acceptors, but that model results are sensitive to other input parameters (inhibition coefficients, maximum rate of reductive dechlorination, biomass concentrations, and groundwater velocity).

Simulation results of enhanced in situ bioremediation using dissolved organic carbon as a reducing agent show that spatial and temporal changes in the dominant TEAP and the subsequent rate of reductive dechlorination are adequately represented with the model. Initial concentrations of Fe(III) and the dechlorinating microbial population influence the simulated time lag observed during the pilot test.

Formulation of the CBC-model for modelling the contaminants and footprints in natural attenuation of BTEX; Maurer M et al.; This paper provides the details of the Coupled Biological and Chemical (CBC) model for representing in situ bioremediation of BTEX. The CBC model contains novel features that allow it to comprehensively track the footprints of BTEX bioremediation, even when the fate of those footprints is confounded by abiotic reactions and complex interactions among different kinds of microorganisms. To achieve this comprehensive tracking of all the footprints, the CBC model contains important new biological features and key abiotic reactions.

The biological module of the CBC-model includes these important new aspects:

- It separates BTEX fermentation from methanogenesis.
- It explicitly includes biomass as a sink for electrons and carbon.

- It has different growth rates for each biomass type.
- It includes inhibition of the different reactions by other electron acceptors and by sulfide toxicants.

The chemical module of the CBC-model includes abiotic reactions that affect the footprints of the biological reactions. In particular, the chemical module describes the precipitation/dissolution of CaCO3, Fe2O3, FeS, FeS2, and S degrees.

Modeling intrinsic bioremediation for interpret observable biogeochemical footprints of BTEX biodegradation: the need for fermentation and abiotic chemical processes; Maurer M et al.; The intrinsic bioremediation of BTEX must be documented by the stoichiometric consumption and production of several other compounds, called 'footprints' of the biodegradation reaction. Although footprints of BTEX biodegradation are easy to identify from reaction stoichiometry, they can be confounded by the stepwise nature of the biodegradation reactions and by several abiotic chemical reactions that also produce or consume the footprints.

In order to track the footprints for BTEX biodegradation, the following reactions need to be considered explicitly:

- Fermentation and methanogenesis as separate processes.
- Precipitation and dissolution of calcite.
- Precipitation and dissolution of amorphous iron monosulfide (FeS).
- Conversion of FeS into the thermodynamically stable pyrite (FeS_2) with loss of sulfide and abiotic formation of H_2.
- Reductive dissolution of solid iron(III) by oxidation of sulfide.

We critically review the research that underlies why these mechanisms must be included and how to describe them quantitatively. A companion manuscript develops and applies a mathematical model that includes these reactions.

Monitoring behaviour of catabolic genes and change of microbial community structures in seawater microcosms during aromatic compound degradation; Sei K et al.; The behaviour of microbial populations responsible for degradation of the aromatic compounds, phenol, benzoate, and salicylate, and changes of microbial community structures in seawater microcosms were analysed quantitatively and qualitatively using MPN-PCR and PCR-DGGE.

The purpose of the study was to investigate the ecology of the entire microbial community during bioremediation. Bacterial populations possessing catechol 1,2-dioxygenase (C12O) DNA were evidently the primary degraders of phenol and benzoate, but others possessing catechol 2,3-dioxygenase (C23O) DNA increased to enhance substrate degradation under high-load conditions when the substrates were present for long periods. However, salicylate degradation was evidently facilitated by specific bacterial populations possessing C23O DNA. PCR-DGGE analyses suggested that bacterial populations already relatively dominant in the original microcosm contributed to phenol degradation.

Bacteria composing a minor fraction of the original population apparently increased and contributed to benzoate degradation. Bacterial populations possessing C23O DNA were responsible for salicylate degradation, however, and different degrading bacteria were evidently selected for, depending on the initial salicylate concentration. Microbial community structure tended to be simplified by aromatic compound degradation.

Thus, microbial monitoring can elucidate the behaviour of bacterial populations responsible for aromatic compound degradation and be used to assess the effects of bioremediation on intact microbial ecosystems.

Molecular diversity of tannic acid degrading bacteria isolated from tannery soil; Chowdhury SP et al.; AIMS: The aim of this study was to enrich and isolate bacteria from a tannery soil that were capable of utilizing tannic acid and gallic acid as sole source of carbon aerobically, and to characterize their diversity in order to identify efficient strains that can be used for tannin bioremediation. METHODS AND RESULTS: Bacterial strains were isolated after enrichment in minimal medium with tannic acid or gallic acid as sole carbon source.

Polymerase chain reaction (PCR) restricted fragment

length polymorphism of 16S rDNA {amplified ribosomal DNA restriction analysis (ARDRA)} and BOX-PCR was used to characterize their diversity. Two strains showing relatively high efficiency in degrading tannic acid and gallic acid were identified on the basis of carbon source utilization pattern (BIOLOG) and 16S rDNA sequence.

Conclusions: Bacterial strains capable of degrading tannic acid and gallic acid could be grouped into six and seven clusters on the basis of ARDRA and BOX-PCR, respectively. i, b, i. On the basis of 16S rDNA sequence, the most efficient isolate degrading tannic acid belonged to Pseudomonas citronellolis, whereas the most efficient gallic acid degrader showed maximum phylogenetic relatedness to P. plecoglossicida. Significance AND IMPACT OF THE STUDY: Aerobic tannic acid degraders such as the two strains isolated in this study can be used for tannin bioremediation, and in the study of genes involved in the production of tannase, an industrially important enzyme.

Effect of root death and decay on dissipation of polycyclic aromatic hydrocarbons in the rhizosphere of yellow sweet clover and tall fescue; Parrish ZD et al.; A 12-mo greenhouse study was conducted to evaluate the contribution of root death and decay on the dissipation of polycyclic aromatic hydrocarbons (PAHs) in rhizosphere soil. The contaminated soil was previously treated by land-farming, but residual PAHs remained after treatment. Tall fescue (Festuca arundinacea Schreb.) and yellow sweet clover (Melilotus officinalis Lam.) were the target plants.

To specifically evaluate the effect of root decay on contaminant dissipation, plants were treated with glyphosate, a broad spectrum herbicide, to induce root decay. Although tall fescue treatments had the highest root and shoot biomass and root surface area, this plant did not result in the highest contaminant degradation rates. Significant differences were noted between treatments for seven PAHs, with the active yellow sweet clover resulting in 60 to 75% degradation of these compounds. Induced root death and decay did not produce a significant enhancement of PAH degradation.

The PAH microbial degrader populations in the vegetated treatments were more than 100 times greater than those in the unvegetated control. The phospholipid fatty acid (PLFA) structural group profile shifted over the growing period, indicating a change in the community structure. In conclusion, phytoremediation was shown to be an effective polishing tool for PAH-affected soil previously subjected to biological treatment.

Ozonation of a landfill leachate: evaluation of toxicity removal and biodegradability improvement; Bila DM et al.; This work shows an evaluation of treatments for the leachate produced at the Gramacho Municipal Landfill in Rio de Janeiro state, Brazil. This leachate has very peculiar characteristics, with a high salinity level and very low biodegradability (BOD(5)/COD of 0.05). A sequence of processes was employed in the treatment of this leachate.

Initially, a physicochemical treatment was used, while the second stage consisted of application of ozone to improve the biodegradability of the leachate. The final stage comprised a biological treatment. The physical-chemical treatment led to COD and DOC removal levels of 40 and 25%, respectively, with the use of Al(2)(SO(4))(3). The sequence of treatments proposed brought good results, with an increase in the BOD(5)/COD ratio from 0.05 to 0.3 after ozonation. The toxicity tests performed using Brachydanio rerio and Poecilia vivipara showed that the toxicity of the leachate had hardly been reduced by ozonation.

These results are in agreement with the fact that, despite the higher BOD(5)/COD ratio, the biological process did not present a good performance. The total average removal levels of COD and DOC achieved using the combined treatment were 73 and 63%, respectively, for an ozone dose of 3.0gL(–1) by the leachate.

Effects of pH amendment on metal working fluid wastewater biological treatment using a defined bacterial consortium; van der Gast CJ et al.; The aim of this study was to determine whether pH amendment of a highly alkaline metal working fluid (MWF) wastewater would improve biological treatment in a bioreactor system following introduction of a bacterial inoculum (comprised of the following strains: Agrobacterium radiobacter, Comamonas testosteroni, Methylobacterium mesophilicum, Microbacterium esteraromaticum, and Microbacterium saperdae).

The pH values tested were 6, 7, 8, and 9. Three replicate batch mode bioreactors inoculated with the bacterial inoculum (plus an abiotic control bioreactor) were operated for each of the four pH conditions. After 14 days, the final mean chemical oxygen demand (COD) reduction at pH 9 was 50 +/– 1.4%; at pH 8, 58 +/– 1.4%; pH 7, 65 +/– 1.0%; and pH 6, 75 +/– 2.7% of the initial COD (approximately 10,000 mg L(-1)), respectively. Interestingly, within 5 days, the pH in all inoculated bioreactors progressed toward pH 8.

However, all abiotic control bioreactors remained at the pH at which they were amended. The fate of the inoculum, determined by denaturing gradient gel electrophoresis (DGGE) and by cluster analysis of the resulting DGGE profiles, revealed that the inocula survived throughout operation of all pH-amended bioreactors. Length-heterogeneity polymerase chain reaction (PCR) was used to track the population dynamics of individual strains. After 7 days of operation, M. esteraromaticum was the most abundant population in all bioreactors, regardless of pH.

From superior adaptation and function to brain dysfunction—the neglect of epigenetic factors; Saugstad LF; With optimal pregnancy conditions (natural, enriched diet which includes fish) African (Digo) infants are 3-4 weeks ahead of European/American infants in sensorimotor terms at birth, and during the first year.

Infants of semi-aquatic sea-gypsies swim before they walk, and have superior visual acuity compared with us. With adverse pregnancy behaviour (fear of fat, a trend to dieting), neglecting the need for brain fat to secure normal brain development and function, we run a risk of dysfunction—death. Sudden Infant Death Syndrome victims have depressed birth weight, lower levels of marine fat in brainstem than controls, and >80 suffer multiple hypoxic episodes prior to death. Depressed birth weight (more than 10% below mean) is seen in learning and behaviour disorders, and a trend towards weights of less than 3kg is increasing, which supports a rise in antenatal sub optimality.

Given marine fat deficiency in pregnancy and infancy, neurons starved for fuel could delay myelination and maturation in the latest developed Frontal Lobes. The phylogenetic oldest Lateral Frontal Lobe System (feed-back mechanism etc.) derived from olfactory bulb-amygdala, which crosses in Anterior Commisure is probably spared, while the Medial Frontal Lobe System derived from Hippocampus-Cingulum and crosses in Corpus Callosum (delayed response task) is most likely affected. The rise in infantile autism (intact vision and hearing) with deficit in delayed response task only, could suggest a deficit in the Medial Frontal Lobe System. The human species is unique; 70% of total energy to the foetus goes to development of the brain, which mainly consists of marine fat. It undergoes pervasive regressive events, before birth, in infancy and at puberty. Minimal retraction of neuronal arborisation is advantageous. Attributable to adverse pregnancy childrearing practice, excessive retraction is likely prenatally and in infancy. Pubertal age affects the fundamental property of nervous tissue, excitability: excessive excitatory drive is seen in early, and a deficiency in late puberty.

Combined quantity management and biological treatment of sludge liquor at Hamburg's wastewater treatment plants—first experience in operation with the Store and Treat process; Laurich F; Store and Treat (SAT) is a new concept for the management of ammonium-rich process waste waters at wastewater treatment plants. It combines the advantages of quantity management and separate biological treatment, whereby both operations are carried out in the same tank. Now the first full-scale application of that method was realized in Hamburg. As first experience shows the process can help to increase nitrogen removal and to reduce energy consumption.

Electrochemical oxidation as a final treatment of synthetic tannery wastewater; Panizza M et al.; Vegetable tannery wastewaters contain high concentrations of organics and other chemicals that inhibit the activity of microorganisms during biological oxidations, so biorefractory organics that are not removed by biological treatment must be eliminated by a tertiary or advanced wastewater treatment. In this paper, the applicability of electrochemical oxidation as a tertiary treatment of a vegetable tannery wastewater was investigated by performing galvanostatic electrolysis using lead dioxide (Ti/PbO2) and mixed titanium and ruthenium oxide (Ti/TiRuO2) as anodes under different experimental conditions.

The experimental results showed that both the electrodes performed complete mineralization of the wastewater. In particular, the oxidation took place on the PbO2 anode by direct electron transfer and indirect oxidation mediated by active chlorine, while it occurred on the Ti/TiRuO2 anode only by indirect oxidation. Furthermore, the Ti/PbO2 gave a somewhat higher oxidation rate than that observed for the Ti/TiRuO2 anode. Although the Ti/TiRuO2 required almost the same energy consumption for complete COD removal, it was more stable and did not release toxic ions, so it was the best candidate for industrial applications. With the Ti/TiRuO2 anode, the rate of tannery wastewater oxidation increased with the current density, pH, and temperature of the solution. These results strongly indicate that electrochemical methods can be applied effectively as a final treatment of vegetable tannery wastewater allowing the complete removal of COD, tannin, and ammonium and deColourization.

Comparison of different advanced oxidation process to reduce toxicity and mineralisation of tannery wastewater; Schrank SG et al.; Many organic compounds contained in wastewater are resistant to conventional chemical and/or biological treatment. Because of this reason different degradation techniques are studied as an alternative to biological and classical physico-chemical processes. Advanced Oxidation Processes (AOPs) probably have developed to become the best options in the near future. AOP while making use of different reaction systems, are all characterised by the same chemical feature: production of OH radicals (*OH).

The versatility of AOPs is also enhanced by the fact that they offer different possibilities for OH radical production, thus allowing them to conform to specific treatment requirements. The main problem with AOPs is their high cost. The application of solar technologies to these processes could help to diminish that problem by reducing the energy consumption required for generating UV radiation. In this work, different AOPs (O_3, TiO_2/UV, Fenton and H_2O_2/UV) were examined to treat tannery wastewater or as a pre-treatment step for improving the biodegradation of tannery wastewater, at different pH and dosage of the chemicals.

Under certain circumstances retardation in biodegradation and/or an increase in toxicity may be observed within these treatment steps. Two different bioassays (Daphnia magna and Vibrio fischeri) have been used for testing the progress of toxicity during the treatment. In parallel other objectives were to analyse and identify organic compounds present in the untreated wastewater and arising degradation products in AOP treated wastewater samples. f, g, j, b. For this purpose substance specific techniques, e.g., gas chromatography-mass spectrometry (GC-MS) in positive electron impact (El(+)) mode and atmospheric pressure ionisation (API) in combination with flow injection analysis (FIA) or liquid chromatography-mass and tandem mass spectrometry (LC-MS or LC-MS-MS) were performed.

Short-term BOD (BODst) as a parameter for on-line monitoring of biological treatment process; Part II: instrumentation of integrated flow injection analysis (FIA) system for BODst estimation; Liu J et al.; An instrument with integrated flow injection analysis (FIA) system has been developed for on-line monitoring a process for conversion of biomass under field condition. The instrument consists of a newly designed biosensor for easy renewal of the bio-receptor without disassembling the sensor, a FIA controller for controlling the analysis operations, and a computer-based data acquisition system for data recording and processing. The instrument performed a sequence operations automatically including preparation of sample in the desired concentration, sample loading, sample injection, signal recording, data processing, and self-cleaning of the system. This makes the instrument being an interesting and promising device for on-line process monitoring.

Short-term BOD (BODst) as a parameter for on-line monitoring of biological treatment process. Part I. A novel design of BOD biosensor for easy renewal of bio-receptor; Liu J et al.; A novel design of a biochemical oxygen demand (BOD) biosensor has been developed for on-line monitoring of easily biodegradable organic compounds in aqueous samples. The biological recognition element of the sensor could be easily renewed by injecting new bacterial paste without disassembling the sensor system. The sensor measurements were carried out in the initial-rate mode using a flow injection (FI) system, resulting in 60 s for one sample analysis followed by a recovery time less than 10 min. The sensor performance achieved showed a wide detection linearity over the range of 5-700 mg BOD5.l(–1) and a generally good agreement between the BOD values estimated by the biosensor and the conventional 5-day test.

Furthermore, the precision test was in the control range (i.e. repeatability < or =/+/–7.5%/, reproducibility < or =/+/–7.3%/). The sensor could be used over 1 week in continuous test, however, the best performance was found within the first 24 h where standard deviation of the sensor response was +/–2.4%. The design of the sensor allows easy and fast renewal of the cells used as sensing elements. Replacement of biological recognition element and calibration of the sensor responses can be performed in a rather simple procedure on a daily regular basis. By using a mixed culture as the bio-receptor, one gets a sensor that reacts to a wide range of substrates. The new sensor construction will thus allow fast and convenient replacement of the bio-receptor and on-line assay of a broad range of substrates. This makes the sensor being an interesting and promising candidate for on-line monitoring of biological treatment process.

Chemical characterisation of natural organic substrates for biological mitigation of acid mine drainage; Gibert O et al.;

The current approach of the biological treatment of acid mine drainage by means of a passive remediation system involves the choice of an appropriate organic substrate as electron donor for sulphate reducers. Nowadays this selection is one of the critical steps in the performance of such treatment, as this depends to a great extent on the degradability of the organic substrate. Thus, a prior characterisation of the organic substrate predicting its biodegradability would be desirable before embarking on an extensive large-scale application.

The aim of this study was to correlate the chemical composition (lignin content) of four different natural organic substrates (compost, sheep and poultry manures, oak leaf) and their capacity to sustain bacterial activity in an attempt to predict biodegradation from chemical characterisation. The results showed that the lower the content of lignin in the organic substrate, the higher its biodegradability and capacity for developing bacterial activity. Of the four organic materials, sheep and poultry manures and oak leaf evolved reducing conditions and sustained active sulphidogenesis, which coupled with the decrease in sulphate concentration indicated bacterial activity. Sheep manure was clearly the most successful organic material as electron donor (sulphate removal >99%), followed by poultry manure and oak leaf (sulphate removal of 80%).

Compost appeared to be too poor in carbon to promote sulphate-reducing bacteria activity by itself. Column experiments emphasised the importance of considering the residence time as a key factor in the performance of continuous systems. With a residence time of 0.73 days, sheep manure did not promote sulphidogenesis. However, extending residence time to 2.4 and 9.0 days resulted in an increase in the sulphate removal to 18% and 27%, respectively.

Landfill leachate treatment by yeast and bacteria based membrane bioreactors; Wichitsathian B et al.; Biological treatment of medium-age landfill leachate was investigated on a membrane bioreactor. The experiments were conducted in two 5-L reactors with immersed hollow fibre microfiltration membranes. One reactor was operated with a mixed bacterial culture termed as bacteria based membrane bioreactor (BMBR) while the other with mixed yeast culture termed as yeast based membrane bioreactor (YMBR).

The leachate was characterized with a chemical oxygen demand (COD) concentration of 7000-9000mg/L, biochemical oxygen demand (5 days) to chemical oxygen demand ratio (BOD5/COD) of 0.35-0.45 and total Kjeldahl nitrogen (TKN) of 1800-2000mg/L. The performance was assessed with and without ammonia stripping. In both the reactors, the average COD and TKN removal efficiency without ammonia stripping ranged between 52-66 and 14-28%, respectively. The performance of both the membrane bioreactors improved with ammonia stripping in terms of both COD (72-76%) and TKN (82-89%) removal efficiency.

Though, the difference in the performance of the BMBR and YMBR was not significant in terms of COD removal, the YMBR showed better removal efficiency in terms of BOD5. The molecular weight cut-off showed that the degradation pathway of the leachate by bacterial and yeast are different. In regard to membrane fouling, the YMBR showed better performance with lower trans-membrane pressure as well as longer operating time. This superior performance of the YMBR could be due to the structure of yeast cells which are larger in size as well as reduced soluble extracellular polymeric substances (EPS) production, which are the main cause of membrane biofouling.

Biological treatment of whitewater in a Labouratory process in order to determine kinetic parameters for model development; Alexandersson T et al.; Implementation of an in-mill biological treatment plant is one solution to the problems associated with closure of whitewater systems. It is, however, important to operate the treatment with low concentration of nutrients in the effluent. The effect on the COD reduction from decreased additions of NH4-N and PO4-P were investigated in three parallel aerobic suspended carrier reactors during treatment at 46 to 48 degrees C of whitewater from a recycled paper mill producing liner and fluting.

In the reference reactor, a COD reduction of 89% was

achieved and 45.6 mg NH4-N/(g COD reduced) and 11.6 mg PO4-P/(g COD reduced) was consumed at an organic load around 20 kg COD/(m3 x d). Reduced additions of NH4-N decreased the COD reduction. Addition of 56% of the consumption of NH4-N in the reference reactor resulted in a COD reduction of 80%. The response from decreased addition of PO4-P was different compared to NH4-N but it could not be determined if this is due to unsuitable experimental design or a different reaction mechanism.

Reducing the addition of PO4-P to 26% of the consumption of PO4-P in the reference reactor, decreased the COD reduction to 83%. The main conclusion from the experiment is: biological treatment has the potential of treating whitewater from recycled paper mills with low effluent nutrient concentrations.

Integrated biological treatment of recalcitrant effluents from pulp mills; Ortega-Clemente A et al.; This work aimed at determining the degree of depuration of a recalcitrant effluent (weak black liquor, WBL) achieved in a series treatment consisting of a first stage methanogenic fluidised bed reactor followed by a second stage aerobic, upflow reactor packed with "biocubes" of Trametes versiColour immobilised onto small cubes of holm oak wood. The mesophilic, lab scale methanogenic fluidised bed reactor contained a microbial consortium immobilised onto granular activated carbon 500 microm average size.

The process removed decreasing amounts of organic matter at decreasing hydraulic retention times (HRT), eventually reaching an average of 50% at 0.5 day HRT. Colour and ligninoid removals also decreased with decreasing HRT. Although the methanogenic fluidised bed reactor provided an effective treatment for the degradable organic matter, important concentrations of recalcitrant organic matter and colour still remained in the anaerobic effluent.

This anaerobic effluent was fed to the aerobic packed bed reactor. Two HRT were tested in this unit, namely 5 and 2.5 days. The reactor averaged an organic matter removal in the range of 32% COD basis, during an experimental run of 95 days. Colour and ligninoid contents were removed in high percentages (69% and 54%, respectively).

There was no significant difference in reactor performance at 5- and 2.5-day HRT. There was a positive correlation between pollutant removal efficiencies and Laccase activity in crude extracts of the reactor liquor. No supplemental soluble carbohydrate was required to sustain the fungus activity and the consistent reactor performance. Overall, the two-stage treatment achieved approximately a 78% removal of the original organic matter of the WBL (COD basis) and ca. 75% of colour and ligninoid contents.

Biostabilization of municipal solid waste; Adani F et al.; A mechanical-biological process for municipal solid waste (MSW) treatment was monitored for one year. Mechanical pre-treatment provided two fractions. The oversize fraction (diameter > 50 mm) (yield of 600 g kg(–1) ww) (46 Mg day(–1)) was used for refuse derived fuel production, after undergoing a mechanical refining processes, because of low moisture content (200–250 g kg(–1)) and high calorific value (2500-2800 kcal kg ww(–1)). The undersize fraction (diameter < 50 mm) (yield 400 g kg(–1) ww) (30 Mg day(–1)) contained about 800 g kg(–1) of the MSW organic matter.

This fraction was biologically treated using an aerobic process with an organic waste fraction from separate collection (77 Mg day(–1)) and recycled stabilized material (62 Mg day(–1)) obtained from end-product sieve (diameter < 20 mm) used as bulking agent. A retention time of three weeks was sufficient to obtain stabilized products in agreement with up-dated rules of the Lombardy Region (North Italy) regarding biostabilization and composting processes. Dynamic Respiration Index (DRI), such as required by both Lombardy Region rules and suggested by the European Community, was chosen in preference to other indices in order to assess the degree of biological stability of the end products.

A mean DRI value of 1164 mg O_2 kg SV(–1) h(–1) was obtained and is in agreement with the proposed limit of 1000+/–200 mg O_2 kg SV(–1) h(–1). Self-heating test, potential biogas production and fermentable volatile solids were also used as parameters to describe the potential impact of treated waste, providing further useful information. Nevertheless, all of these methods revealed analytical or interpretative limits. A complete mass balance of the biological treatment section showed that, from a net input of 107 Mg day(–1), only 250 g kg(–1) (27 Mg day(–1)) of the waste needed to be landfilled, with 750 g kg(–1) (80 Mg day(–1)) being lost as CO_2 and H_2O.

Metal solubilization from metal-containing solid materials by cyanogenic Chromobacterium violaceum; Faramarzi MA et al.; Different cyanogenic bacterial strains (Chromobacterium violaceum, Pseudomonas fluorescens, Bacillus megaterium) were cultivated under cyanide-forming conditions in the presence of metal-containing solids such as nickel powder or electronic scrap.

All microorganisms were able to form water-soluble metal cyanides, however, with different efficiencies. C. violaceum was able to mobilize nickel as tetracyanonickelate {Ni(CN)4(2–)} from fine-grained nickel powder. Gold was microbially solubilized as dicyanaoaurate {Au(CN)2–} from electronic waste.

Additionally, cyanide-complexed copper was detected during biological treatment of shredded printed circuit boards scrap. Regarding the formation of tetracyanonickelate, C. violaceum was more effective than P. fluorescens or B. megaterium. Besides a few previous reports on gold solubilization from gold-containing ores or native gold by C. violaceum, the findings demonstrate for the first time the microbial mobilization of metals other than gold from solid materials and represent a novel type of microbial metal mobilization based on the ability of certain microbes to form HCN.

The results might have the potential for industrial applications (biorecovery, bioremediation) regarding the treatment of metal-containing solids since metal cyanides can easily be separated by chromatographic means and be recovered by sorption onto activated carbon.

Fate and stability of 14C-labeled 2,4,6-trinitrotoluene in contaminated soil following microbial bioremediation processes; Weiss M et al.; Biological treatment of 2,4,6–trinitrotoluene (TNT) in soil rarely results in complete mineralization of the parent compound. More often, the largest proportion of the TNT carbon is incorporated into the soil organic matrix. Therefore, we evaluated the stability of nonextractable residues from various bioremediation processes of 14C-TNT in soils. The extractable amounts of the residual radioactivity varied between 7 and 33% and thus the nonextractable amount between 93 and 67% (3-15% in fulvic acids, 26-46% in humic acids, and 27-44% in the humin fraction). The residue-containing soils were Analysed for the release of radioactivity after treatment by physical (freeze and thaw, grinding of soil, and steam extraction), chemical (acid rain and addition of metal complexing agent), and biological methods (addition of compost, white rot fungi, radical-generating enzymes, and germination of plants).

Freeze and thaw treatment and grinding of the soil did not alter the partitioning of the label significantly. Steam extraction and acid rain extraction increased the water extractability to 11 to 29% and to 51.6% in the native TNT-contaminated soil. The addition of ethylenediamine-tetraacetate (EDTA) increased the extractability from 7 to 12%. d, h, a, c, j, b. After biological treatment, only slightly increased extractability (<10%) was observed. No increase of extractable TNT or known metabolites was observed with any of the treatments. Thus, under the treatment conditions applied in this study, the residues formed during microbial transformation of TNT may be biogenic residues with low mobilization potential and low hazardous impact.

Biotreatment of wastewater using aquatic invertebrates, Daphnia magna and Paramecium caudatum; Shiny KJ et al.; A number of major changes have occurred over the past few years, which give cause for a re-examination of conventional wastewater treatment methods.

Among these are growing problems of worldwide energy and food shortages and nutrients not removed by conventional secondary processes causing algal blooms and other problems in the receiving waters.

The global increase in wastewater calls for innovative low cost technology approaches to its recycling. Biotreatment systems, utilizing living organisms are receiving growing attention since they are ecologically sound, cheap and applicable in areas without land constraints. Filter feeders (both invertebrates and vertebrates) are promising in this area since they can remove suspended organic matter and bacteria, even in the size range of microns.

In the present study biological treatment of municipal wastewater using two invertebrates—Paramecium caudatum, a protozoan and Daphnia magna, a cladoceran was investigated. Analysis at pre-experimental and post-experimental stages revealed the potential of these species in abatement of water pollution. D. magna was more efficient than P. caudatum in Labouratory-scale studies.

Fe(III)-EDTA complex abatement using a catechol driven Fenton reaction combined with a biological treatment; Oviedo C et al.; A combined chemical oxidation (catechol-driven Fenton reaction) followed by a biological treatment was used to degrade Fe(III)-EDTA (1.34 mM). The chemical treatment was inspired in fungal non-enzymatic wood rot mechanisms that use dihydroxybenzens in order to promote the Fenton reaction to breakdown wood structures.

This chemical pre-oxidation used catechol (50 microM) and H202 (20 mM) and the reaction products were identified by GC-MS. In addition, a biological treatment was coupled using the baker's yeast Saccharomyces cerevisiae. The combined chemical biological treatment achieved 100% EDTA degradation, 68 % total organic carbon removal and 90% iron removal.

The World Federation of Societies of Biological Psychiatry (WFSBP) guidelines for the biological treatment of bipolar disorders, part III: maintenance treatment; Grunze H et al.; As with the two preceding guidelines of this series, these practice guidelines for the pharmacological maintenance treatment of bipolar disorder were developed by an international task force of the World Federation of Societies of Biological Psychiatry (WFSBP).

Their purpose is to supply a systematic overview of all scientific evidence relating to maintenance treatment. The data used for these guidelines were extracted from a MEDLINE and EMBASE search, from recent proceedings from key conferences and various national and international treatment guidelines.

The scientific justification of support for particular treatments was categorised into four levels of evidence (A-D). As these guidelines are intended for clinical use, the scientific evidence was not only graded, but also reviewed by the experts of the task force to ensure practicality.

Evaluation of virological quality of sewage from four biological treatment plants by a nested-PCR technique; Komninou G et al.; In order to determine the virological quality of sewage from four biological treatment plants in Greece (two in the city of Athens and two in the city of Patras), 92 raw sewage samples were analysed for the presence of enteroviruses and adenoviruses during the period from October 2000 to February 2003.

A nested-PCR method was used in order to increase the sensitivity of virus detection. Enteroviruses were detected in 43 samples (47%) and adenoviruses in 75 samples (81.5%) of raw sewage by nested PCR. The more frequent isolation of adenoviruses in raw sewage indicated their stability as virological indicators of the pollution of the environment and their increased persistence in sewage.

Hydration properties of eco-cement pastes from waste sludge ash clinkers; Lin KL et al.; Three types of hydraulic cements have been developed by incorporating sludge ash from a primary sewage treatment plant and a water purification plant, as well as slag from steelworks (ferrate), as a partial replacement for clay, silica, alumina, and iron oxide in raw cement meal. The raw meal for the pre-determined recipes was prepared by heating it to 1400 degrees C for 6 hr in a clinkerization process, using a simulated incinerator and smelter.

The major components of ordinary Portland cement, C_3S, C_2S, C_3A, and C_4AF, were all found in the clinkers. Of the three types of eco-cements, the eco-cement A paste was most similar to ordinary Portland cement in terms of composition and compressive strength development, while the eco-cement B paste showed early strength development.

The differential thermal analysis species analyses indicated that the hydrates in the eco-cement pastes were mainly calcium hydroxide and CSH gels, like those found in ordinary Portland cement paste. Moreover, the degree of hydration, as determined by nuclear magnetic resonance, increased in all eco-cement pastes with an increasing curing age. The results indicate that it indeed is feasible to use sludge ash and ferrate to replace up to 20% of the mineral components of raw materials for cement. Aquatic ecosystem as a bioreactor: water purification and some other functions; Ostroumov SA; A fundamental concept is proposed of aquatic ecosystem as a bioreactor that carries out the function of water purification in natural water bodies and streams.

The ecosystem as a bioreactor has the following characteristic attributes:

- It is a large-scale (large-volume) bioreactor.
- It is a diversified (in terms of the number of taxa and the scope of functional activities) bioreactor.
- It possesses a broad range of biocatalytic (chemical-transforming and degrading) capabilities.

New experimental data on xenobiotics-induced inhibition of the functioning of the molluscs Unio tu- midus, U. pictorum, M. galloprovincialis and Limnaea stagnalis emphasize the potential ecological hazard from sublethal concentrations of pollutants (including those exemplified by synthetic surfactants and detergents).

Detection of enteric viruses and bacterial indicators in German environmental waters; Pusch D et al.; A German mining lake and the supplying surface waters, which are located downstream of a sewage plant, were examined regarding their microbiological and virological quality. Between October and September 2008, specific PCR methods were used to determine the occurrence of enteric viruses in 123 water specimens drawn at different sites downstream of the waste water treatment plant and in 9 samples from the sewage plant influent.

Detection rates in sewage plant effluents and surface water samples depended on sampling sites and were: 29–76% for enterovirus (EntV), 24–42% (astrovirus, AstV), 15-53% (norovirus, NV), 3-24% (rotavirus, RoV), 5–20% (hepatitis A virus, HAV) and 20% (adenovirus, AdV). AstV genome load of selected samples was between $3.7 \times 10(3)$ to $1.2 \times 10(8)$ genome equivalents per liter (gen.equ./l), depending on sampling location; NV average genome load ranged from $1.8 \times 10(4)$ to $9.7 \times 10(5)$ gen.equ./l.

Cell culture methods showed that three out of 18 PCR positive samples contained infectious EntV. Even though microbiolical parameters such as Escherichia coli, enterococci and coliphages indicated acceptable microbiological water quality, the virological data of this study suggest the possibility that surface waters may be a source for enteric viral infections.

Studies on the effect of humic acids and phenol on adsorption-ultrafiltration process performance; Mozia S et al.; Application of pressure-driven membrane processes, such as ultrafiltration (UF) and microfiltration (MF) for surface water treatment have become very popular during last decades. Membrane fouling by humic substances (HS) is one of the major limiting factors in these processes. In order to alleviate the unFavourable effects of the presence of HS in the feed on the process performance UF and MF are often combined with adsorption on powdered activated carbon (PAC).

The main goal of the present study was to evaluate the effect of humic acid (HA) on membrane fouling during UF. Moreover, the effect of PAC addition to the feed on UF process, especially on flux decline was determined. The applicability of the adsorption-ultrafiltration (PAC/UF) system to purification of water containing low (phenol) and high molecular (HA) was also investigated. Three different polymer UF membranes, prepared from polysulfone (PSF), cellulose acetate (CA) or polyacrylonitrile (PAN) were applied.

It was found that the membranes prepared from PSF and CA are very susceptible to fouling caused by HA. The permeate flux decreased for ca. 50% during UF of HA solution through the PSF membrane and for ca. 45%–through the CA membrane. In the case of the PAN membrane, a negligible effect of HA on the flux was observed. On the basis of the FTIR spectra it was found that the drop in the permeate flux through these membranes may result from interactions between the negatively charged functional groups present on the membrane surface, such as carboxyl groups (CA) and sulfone groups (PSF) with HA, which results in coating of the membrane surface with HA.

When PAC was added to the feed containing HA, the permeate flux through the CA and PAN membranes was maintained on a practically unchanged level. However, in case of the PSF membrane, a 50% drop in the permeate flux in comparison with the flux value, when process was conducted without PAC addition was observed. That was supposed to be due to attractive forces among hydrophobic PAC particles, HA molecules and PSF membrane surface. The performed studies showed that the application of PAC/UF system was very effective in the removal of organic substances having both, low and high molecular weights. The role of PAC suspended in a feed in the PAC/UF system is the adsorption of low molecular organic compounds, which cannot be removed by UF alone.

Optimization of the extended terminal subfluidization wash (ETSW) filter backwashing procedure; Amburgey JE; The increased passage of particles and microorganisms through granular media filters immediately following backwashing is a common problem known to the water treatment community as filter"ripening"or maturation. While several strategies have been developed over the years to reduce the impact of this vulnerable period of the filtration cycle on finished water quality, this research involves a recently developed filter backwashing strategy called the extended terminal subfluidization wash (ETSW). ETSW is a method of terminating the backwash cycle with a subfluidization wash for a period of time sufficient to pass one theoretical filter-volume of water upward through the filter. ETSW was shown to remove significantly greater quantities of backwash remnant particles thereby reducing the magnitude of filter ripening turbidity and particle count spikes.

Optimum ETSW flow rates were determined for deep-bed anthracite and granular activated carbon filters herein by monitoring filter effluent turbidities and particle counts during the filter ripening period. Optimality of the coagulation process was also shown to influence the magnitude of filter ripening particle passage. ETSW was found to be equally effective for biological and conventional deep-bed anthracite filters.

An improvement on the power law for the description of particle size distributions in potable water treatment; Ceronio AD et al.; In potable water treatment, the use of the power law to describe particle size distributions (PSDs) in particle counting practice is common. The power law is popular because it allows the reduction of numerous data bits to two meaningful parameters that completely describe the size distribution characteristics of a particle suspension. The model is however flawed. This paper presents the further development of an improved model (the variable-beta model) first proposed by Lawler (1997). Both the power law model and variable-beta model are used to describe the PSDs of a large number of potable water treatment samples taken from full-scale plants and the resulting correlations are compared. The findings from the comparison of data reduction methodologies support the argument that the variable-beta model is fundamentally more correct than the power law model and consequently describes the PSDs better.

Melatonin improved the disturbances in hepatic prooxidant and antioxidant balance and hepatotoxicity induced by a high cholesterol diet in C57BL/6J mice; Balkan J et al.; We examined the effect of melatonin in prooxidant and antioxidant state in the liver of C57BL/6J mice fed on a high cholesterol (HC) diet. Mice were fed with normal mice chow containing 1.5% cholesterol and 0.5% cholic acid for 4 months without and with melatonin (10 mg/L in drinking water) treatment. HC diet was observed to increase malondialdehyde (MDA) and diene conjugate (DC) levels in the liver.

This diet lowered glutathione (GSH), alpha-tocopherol, and total ascorbic acid levels as well as glutathione peroxidase (GSH-Px) and glutathione transferase (GST) activities in the liver, but hepatic superoxide dismutase (SOD) activity remained unchanged. Although melatonin treatment did not affect these parameters in mice fed a normal diet, it reduced hepatic MDA and DC levels in mice fed an HC diet. Hepatic alpha-tocopherol and ascorbic acid levels increased, but hepatic GSH levels remained unchanged in the melatonin-treated HC group as compared to the HC group. Melatonin treatment was found to increase liver GSH-Px and GST activities in mice fed an HC diet.

However, SOD activity did not alter in the liver of hypercholesterolemic mice following melatonin treatment. In addition, the histopathological lesions observed in the cholesterol-plus-melatonin group were less severe than those seen in the cholesterol group. According to these observations, we can say that melatonin treatment has an amelicrating effect on the disturbances in prooxidant and antioxidant balance and histopathological lesions in the liver of mice following cholesterol feeding.

Comparative study of different surface decontaminants on chicken quality; Sinhamahapatra M et al: (1)A comparative study on the effect of different surface decontaminants: hot water at 70 degrees C for one minute; 2% lactic acid for 30 s; 1200 p.p.m. acidified sodium chlorite (ASC) solution for 5 s and 50 p.p.m. chlorine solution for 5 min in the form of dips and sprays on the surface of dressed broilers for 0, 24 and 48 h of storage was conducted. (2) The variables studied were, total plate count (TPC), presumptive coliform count (PCC), pH and extract release volume (ERV). All treatments reduced TPC and PCC. (3) Lactic acid dip and hot water dip were the most effective for reducing TPC (1.36 and 1.28 log/cm^2, respectively) with no significant difference between them. (3) ASC and hot water in dip could diminish PCC (1.37 and 1.34 log/cm^2, respectively) and did not vary significantly. (4) No treatment affected muscle pH, water holding capacity (WHC), ERV, appearance, smell, tenderness and overall acceptability of treated broilers significantly. (5) Hot water treatment is the cheapest, most convenient and simplest decontamination technique for hygienic and wholesome poultry production. It changed very little in distribution system. (2) Traditional treatment process can remove MAP efficiently 34.0%-83.7% of MAP can be removed in the process. (3) Generally in source water and water treatment process, there was no evident difference between AOC(potential), AOC(P) and AOC(native). So AOC was the most important factor of bacterial growth. In the main distribution pipe, the concentration of AOC(potential) and AOC(P) were 2-8.7 times of the concentration of AOC(native). So phosphorus limited bacterial regrowth.

Performance of a constructed wetland treating intensive shrimp aquaculture wastewater under high hydraulic loading rate; Lin YF et al.; A water treatment unit, mainly consisting of free water surface (FWS) and subsurface flow (SF) constructed wetland cells, was integrated into a commercial-scale recirculating aquaculture system for intensive shrimp culture. This study investigated performance of the treatment wetlands for controlling water quality.

The results showed that the FWS-SF cells effectively removed total suspended solids (55–66%), 5-day biochemical oxygen demand (37–54%), total ammonia (64–66%) and nitrite (83–94%) from the recirculating water under high hydraulic loading rates (1.57–1.95m/day). This led to a water quality that was suitable for shrimp culture and effluent that always satisfied the discharge standards. The area ratios of wetlands to culture tank being demonstrated (0.43) and calculated (0.096) in this study were both significantly lower than the reported values. Accordingly, a constructed wetland was technically and economically feasible for managing water quality of an intensive aquaculture system.

RULES

Water, sanitation, and hygiene interventions to reduce diarrhoea in less developed countries: a systematic review and meta-analysis; Fewtrell L et al.; Many studies have reported the results of interventions to reduce illness through improvements in drinking water, sanitation facilities, and hygiene practices in less developed countries. There has, however, been no formal systematic review and meta-analysis comparing the evidence of the relative effectiveness of these interventions. We developed a comprehensive search strategy designed to identify all peer-reviewed articles, in any language, that presented water, sanitation, or hygiene interventions.

We examined only those articles with specific measurement of diarrhoea morbidity as a health outcome in non-outbreak conditions. We screened the titles and, where necessary, the abstracts of 2120 publications. 46 studies were judged to contain relevant evidence and were reviewed in detail. Data were extracted from these studies and pooled by meta-analysis to provide summary estimates of the effectiveness of each type of intervention. All of the interventions studied were found to reduce significantly the risks of diarrhoeal illness.

Most of the interventions had a similar degree of impact on diarrhoeal illness, with the relative risk estimates from the overall meta-analyses ranging between 0.63 and 0.75. The results generally agree with those from previous reviews, but water quality interventions (point-of-use water treatment) were found to be more effective than previously thought, and multiple interventions (consisting of combined water, sanitation, and hygiene measures) were not more effective than interventions with a single focus. There is some evidence of publication bias in the findings from the hygiene and water treatment interventions.

Efficacy of electrolyzed oxidizing water for the microbial safety and quality of eggs; Bialka KL et al.; During commercial processing, eggs are washed in an alkaline detergent and then rinsed with chlorine to reduce dirt, debris, and microorganism levels. The alkaline and acidic fractions of electrolyzed oxidizing (EO) water have the ability to fit into the 2-step commercial egg washing process easily if proven to be effective. Therefore, the efficacy of EO water to decontaminate

Salmonella Enteritidis and Escherichia coli K12 on artificially inoculated shell eggs was investigated.

For the in vitro study, eggs were soaked in alkaline EO water followed by soaking in acidic EO water at various temperatures and times. Treated eggs showed a reduction in population between > or = 0.6 to > or =2.6 log10 cfu/g of shell for S. Enteritidis and > or =0.9 and > or =2.6 log10 for E. coli K12. Log10 reductions of 1.7 and 2.0 for S. Enteritidis and E. coli K12, respectively, were observed for typical commercial detergent-sanitizer treatments, whereas log10 reductions of > or =2.1 and > or =2.3 for S. Enteritidis and E. coli K12, respectively, were achieved using the EO water treatment. For the pilot-scale study, both fractions of EO water were compared with the detergent-sanitizer treatment using E. coli K12. Log10 reductions of > or = 2.98 and > or = 2.91 were found using the EO water treatment and the detergent-sanitizer treatment, respectively. d, e, h,k. The effects of 2 treatments on egg quality were investigated. EO water and the detergent-sanitizer treatments did not significantly affect albumen height or eggshell strength; however, there were significant affects on cuticle presence. These results indicate that EO water has the potential to be used as a sanitizing agent for the egg washing process.

Polycyclic musks in the Ruhr catchment area-transport, discharges of waste water, and transformations of HHCB, AHTN and HHCB-lactone; Bester K; The polycyclic musk fragrance compounds HHCB (1,3,4,6,7,8-hexahydro-4,6,6,7,8,8-hexamethylcyclopenta-(g)-2-benzopyran; trade name, e.g. galaxolide{registered sign}) and AHTN (7-acetyl-1,1,3,4,4,6-hexamethyl-1,2,3,4-tetrahydronaphthalene, trade name, e.g. tonalide{registered sign}) and the transformation product of HHCB (HHCB-lactone) were analysed in surface water samples and sewage treatment plants (STP) effluents in the Ruhr megalopolis. The STPs were the dominant source for these pollutants. In the part of the river where the drinking water is extracted from the river, about 60 ng L(-1) HHCB, 10 ng L(–1) AHTN and 20-30 ng L(–1) HHCB-lactone were found as typical riverine concentrations, while none of the compounds were detected near the spring of the river. On the other hand sewage treatment plant effluents exhibited concentrations up to 600 ng L(–1). The STP's effluent resulted in elevated concentrations in some parts of the river and in the lakes into which they discharge. As some of the plants emit HHCB-lactone with a significantly changed enantiomeric pattern, biotransformation of HHCB to HHCB-lactone occurs in some waste water treatment plants operating with activated sludge. In those parts of the river where no relevant discharges of waste water or fresh water takes place neither the concentration nor the pattern changes significantly. This holds true especially for the HHCB versus HHCB-lactone ratios which indicates degradation less than 15% of the HHCB inventory in the river Ruhr itself. In other rivers, such as the Rhine, higher levels of HHCB-lactone in comparison to HHCB were detected (ratio 1 {ratio} 1).

Present and future control of cryptosporidiosis in humans and animals; Jenkins MC; Although water treatment processes to remove Cryptosporidium are improving and detection methods for identifying the parasite in water are becoming more sensitive, outbreaks of cryptosporidiosis continue in the human population.

Animals, especially dairy calves, often become infected as C. parvum oocysts are present in high numbers, remain viable for long periods of time after excretion and no disinfectants, prophylactic or therapeutic reagents exist. Vaccination against C. parvum is being attempted in ruminants for the purpose of generating hyperimmune colostrum containing antibodies that may be effective in passive immunotherapy against cryptosporidiosis in the young. A number of recombinant C. parvum surface or internal antigens have been expressed by DNA-cloning technology. Immune colostrum specific for several recombinant C. parvum proteins have demonstrated efficacy in murine and ruminant models against cryptosporidiosis. The P23 and CP15 antigens appear to be the most promising candidates for vaccine development. Recent studies have demonstrated efficacy of the drug nitazoxanide against C.

parvum infection in humans. In the near future, control of this parasitic disease in humans and animals will rely on a combination of passive immunotherapy and selective drug treatment.

Surimi wash water treatment for protein recovery: effect of chitosan-alginate complex concentration and treatment time on protein adsorption; Wibowo S et al.; Chitosan (Chi), a protein recovery agent for the treatment of aqueous food processing streams, appears to work by mechanical entrapment and electrostatic interaction of chitosan amino groups with anionic groups on proteins. Chitosan effectiveness for recovering soluble proteins from surimi wash water (SWW) is increased by complexation with alginate (Alg) and by adjusting complex concentration and treatment time.

Flocculation at 20 degrees C with Chi-Alg at a 0.2 mixing ratio added as 20, 40, 100 and 150 mg/L SWW was aided by 5 min agitation at 130 rpm and then held at the same temperature for 30 min, 1 and 24 h. Turbidity measurements, protein determinations and qualitative FTIR analysis confirmed SWW protein adsorption which depended on Chi-Alg concentration and reaction time while turbidity reduction was affected by concentration only. No differences ($p<0.05$) in protein adsorption were found between 1 and 24 h. Using 100 mg Chi-Alg complex/L SWW for 1 h achieved 83% protein adsorption and 97% turbidity reduction.

Reuse of fresh water sludge in cement making; Pan R et al.; With the increasing demand for high quality water, a large quantity of chemical agent must be added in the water purification process, which in turn generates enormous amount of fresh water sludge. Of all the options for sludge disposal, sludge reuse has been considered most economical and environmentally sound. This study evaluated the possibility of incorporating fresh water sludge in the making of Portland cement through the sintering process. The goal was to search for the optimal condition to maximize the replacement of clay with the fresh water sludge. Characteristics of fresh water sludge were collected and Analysed. The analysis showed that water source and water treatment process dominate th characteristics, particularly the chemical composition of the fresh water sludge.

The fresh water sludge was mixed with the cement clay in various percentages, from 0% to 100%, as raw material for cement-making. The effects of its addition on the sintering condition and the quality of cement were evaluated. The analysis of the clinkers showed that the addition of the fresh water sludge did not change the phase form and the f-CaO content of the cement. The compressive strength of the masonry increased with the increasing addition of fresh water sludge. All cement products made from various replacement ratios met the Chinese National Standard of first degree Portland cement.

Acidic and hydrogen peroxide treatment of poly-aluminum chloride (PACL) sludge from water treatment; Kwon JH et al.; The water treatment sludge including coagulants cannot be easily removed by conventional dewatering methods. The possibility of hydrogen peroxide (H2O2) oxidation as a pretreatment to enhance the dewaterability of polyaluminum chloride (PACl) sludge from water works was investigated. H_2O_2 treatment alone was not effective but H2O2 treatment under acidic condition significantly reduced both the cake water content and specific resistance to filtration (SRF), indicating the enhancement of dewaterability and filterability.

The filterability after acid/H2O2 treatment was comparable to polymer conditioning and even more dewatered cake than polymer conditioning was produced. By H202 combined with sulfuric acid (H2SO4), leached iron caused Fenton's reaction, which showed a potential to significantly reduce the amount of solids mass and to produce more compact cake with higher filterability.

Although polyphosphate has been found in many organisms, only few of PAOs were isolated, cultured and identified. Culture medium formulation is the key to isolate PAOs and to study the microbial accumulation of polyphosphate, and the competition of glycogen-accumulating organisms (GAOs) with PAOs for carbon resources is one of the reasons of low EBPR efficiency. Modern scientific methods such as fluorescent in situ hybridization, confocal laser scanning microscope, microautoradiography, and in vivo NMR spectroscopy, provided powerful tools to Analyse PAO species composition, spatial structure and functional properties under field conditions. The knowledge of PAO is valuable to enhance the P removal efficiency in water treatment plant, and to improve our understanding on P transformation and transferring in environment.

Particles in filter effluent: the roles of deposition and detachment; Kim J et al.; Particles in the effluent of granular media filters can be classified as influent particles that were never removed or as particles that detached after prior deposition. To determine the effects of particle size, filter media depth and filter run duration on the relative fraction of each class, Labouratory experiments were performed using suspensions of four sizes of polystyrene particles (0.2, 1.2, 2.5, and 4.0 microm diameters) that were destabilized with 0.04 M calcium chloride and continuously supplied to filters after flocculation. To investigate particle attachment alone, three sizes (1.4, 4.0, and 9 microm) of fluorescent microspheres (FM) were periodically pulse injected immediately ahead of the filter media. Detachmentwas assessed as the difference between net removal (particle counts) and deposition (FM counts). FM deposition followed theory, while results show that particle detachment was significant from an early phase of filtration (100 minutes). The detached fraction of effluent particles increased with particle size (1 to 12 microm range) and filter depth. These model system results suggest that detachment plays a significant role in the origin of filter effluent particles in full-scale water treatment systems.

Oxidation of the cyanobacterial hepatotoxin microcystin-LR by chlorine dioxide: reaction kinetics, characterization, and toxicity of reaction products; Kull TP et al.; Cyanobacteria are known producers of cytotoxins, hepatotoxins, and neurotoxins. The main toxins are microcystins, cyclic heptapeptide hepatotoxins, produced by strains of several cyanobacterial genera frequently found in eutrophied freshwaters. Due to the acute and chronic toxicity of microcystins, successful removal of these toxins in drinking water treatment processes is of increasing concern.

In the present work the kinetics of microcystin-LR (MC-LR) oxidation by chlorine dioxide (ClO_2) was studied with UV-spectrometry and high performance liquid chromatography (HPLC).

Characterization of reaction products was performed with mass spectrometric (MS) analysis, while the toxicity of reaction products was tested with a protein phosphatase inhibition assay (PPIA). The main reaction products formed, dihydroxy isomers of MC-LR as identified by MS, were nontoxic according to the PPIA. The overall rate constant k for the reaction between MC-LR and ClO2 at 293 K and pH 5.65 was modest, k = 1.24 M(-1) s(-1), suggesting that ClO2 is not a suitable oxidant for the degradation of microcystins in drinking water treatment processes.

Remote site production of sterile purified water from available surface water; Taylor MA et al.; A water purification and sterilization device was tested for its functional capabilities. Challenge water consisting of potable water augmented with bacteria, endotoxin, virus, suspended solids, and dissociable ions (sodium chloride, lead or arsenic salts) was passed through the device. The product water quality attributes were Analysed. The device demonstrated reduction in bacteria of >7 logs, endotoxin was reduced by >4 logs, virus was reduced by >4 logs, and dissociable ions were reduced by >3 logs. The product water of the device met the limits for a range of chemical entities specified by the United States Pharmacopeia and Association for the Advancement of Medical Instrumentation.

The product water met the quality attributes of Sterile Water for Injection, USP, Sterile Purified Water, USP, and the Water for Dialysis. The device provides a logistical advantage in reducing the weight of transport of packaged water by 83% and the cube by 67%. It operates manually by gravity and is disposable after a single use. The device provides an effective alternative to the transport and use of packaged sterile water in remote locations by production of sterile water at the point-of-need using available water.

It also is capable of producing safe drinking water following the production of clinical waters. This device has been cleared by the US Food and Drug Administration for production of three liters Sterile Purified Water, USP from Environmental Protection Agency (EPA) grade drinking water.

The effect of fruit age on seed germinability of a heterocarpic species, Atriplex sagittata; Mandak B et al.; Atriplex sagittata is an annual heterocarpic plant that produces three different fruit types (termed A, B, and C). To examine the consequence of heterocarpy on germinability patterns over long time periods, we compared seed germinability of different fruit types that had been stored for up to 8 years. While germinability of non-dormant type C fruits in distilled water was high (up to 100 %) in the first 2 years, it rapidly decreased over time. Dormant fruit types A and B showed increased germinability up to 7 years, though loss of germinability was lower for type B than for type A fruits.

Eight-year-old fruits of all types had significantly lower germinability than younger fruits, probably due to loss of viability. Heterocarpy, therefore, ensures that emergence rates for seedlings of A. sagittata will be maintained over relatively long periods, even in years of strong disturbance when all adult plants may be destroyed. The experiment further showed that germinability of all fruit types in high concentrations of salt, as compared with water treatment, changed over the course of 8 years. Whilst dormant types (A and B) of A. sagittata show increased germinability with age of the seed in water treatment, they significantly lose germinability over time with salinity treatment. Type C fruit was not influenced by salt in the first year, but germinability rapidly decreased with time. It follows that the species is able to germinate under high salt concentration in the first year, but this advantage gradually disappears.

Drinking water treatment processes for removal of Cryptosporidium and Giardia; Betancourt WQ et al.; Major waterborne cryptosporidiosis and giardiasis outbreaks associated with contaminated drinking water have been linked to evidence of suboptimal treatment. Cryptosporidium parvum oocysts are particularly more resistant than Giardia lamblia cysts to removal and inactivation by conventional water treatment (coagulation, sedimentation, filtration and chlorine disinfection); therefore, extensive research has been focused on the optimization of treatment processes and application of new technologies to reduce concentrations of viable/infectious oocysts to a level that prevents disease.

The majority of the data on the performance of treatment processes to remove cysts and oocysts from drinking water have been obtained from pilot-tests, with a few studies performed in full-scale conventional water treatment plants. These studies have demonstrated that protozoan cyst removal throughout all stages of the conventional treatment is largely influenced by the effectiveness of coagulation pretreatment, which along with clarification constitutes the first treatment barrier against protozoan breakthrough.

Physical removal of waterborne Crytosporidium oocysts and Giardia cysts is ultimately achieved by properly functioning conventional filters, providing that effective pretreatment of the water is applied. Disinfection by chemical or physical methods is finally required to inactivate/remove the infectious life stages of these organisms.

The effectiveness of conventional (chlorination) and alternative (chlorine dioxide, ozonation and ultra violet {UV} irradiation) disinfection procedures for inactivation of Cryptosporidium has been the focus of much research due to the recalcitrant nature of waterborne oocysts to disinfectants. This paper provides technical information on conventional and alternative drinking water treatment technologies for removal and inactivation of the protozoan parasites Cryptosporidium and Giardia.

Monitoring of COD as an organic indicator in waste water and treated effluent by fluorescence excitation-emission (FEEM) matrix characterization; Lee S et al.; The fluorescence excitation-emission matrices (FEEM) of domestic waste water, treated effluent of a waste water treatment plant and receiving river water were Analysed to select wavelengths for the monitoring of organic contents as COD. Excitation/emission wavelengths of 220/350 nm and 270/350 nm for protein-like fluorescence and 240/450 nm and 340/450 nm for humic-like fluorescence were suggested as fluorescence peak emitting wavelength pairs, respectively.

Without any pre-treatment, the protein-like fluorescence peaks showed better correlation between COD values and fluorescence intensities than the humic-like fluorescence peaks. No enhanced correlation was observed by removing the suspended solids from the samples using filtration. However, statistical multiple regression methods, using the fluorescence intensities from each peak and the light scattering intensity at 633 nm as variables, resulted in an enhanced correlation, with $r2 > 0.9$ for the measured and predicted COD values.

Using artificial streams to assess the effects of metal-mining effluent on the life cycle of the freshwater midge (Chironomus tentans) in situ; Hruska KA et al.; In 2002, we developed an in situ life-cycle bioassay with Chironomus tentans in artificial streams to evaluate the effects of a complex metal mine effluent under ambient environmental conditions. The bioassay was tested in the field using effluent from the Copper Cliff Waste Water Treatment Plant at INCO (Sudbury, ON, Canada). Chironomus tentans were exposed throughout the life cycle to 45% Copper Cliff effluent, which is the average effluent concentration measured in Junction Creek (ON, Canada), the natural receiving environment.

Chironomus tentans in the effluent treatment exhibited reduced survival ($p = 0.001$), reduced total emergence ($p = 0.001$), increased time-to-emergence ($p = 0.001$), and reduced hatching success ($p = 0.001$) relative to animals in the reference water treatment. c, l, c, g, i,k. Chironomus tentans in the effluent treatment were not significantly different from the reference in terms of growth, sex ratio, number of egg cases/female, and number of eggs/egg case. This research showed how a life-cycle bioassay could be used in situ to assess metal mine effluent effects on a benthic invertebrate.

Identification and understanding of fouling in low-pressure membrane (MF/UF) filtration by natural organic matter (NOM); Lee N et al.; An understanding of natural organic matter (NOM) as a membrane foulant and the Behaviour of NOM components in low-pressure membrane fouling are needed to provide a basis for appropriate selection and operation of membrane technology for drinking water treatment. Fouling by NOM was investigated by employing several innovative chemical and morphological analyses. Source (feed) waters with a high hydrophilic (HPI) fraction content of NOM resulted in significant flux decline. Macromolecules of a relatively hydrophilic character (e.g.

Polysaccharides) were effectively rejected by low-pressure membranes, suggesting that macromolecular compounds and/or colloidal organic matter in the hydrophilic NOM fraction may be a problematic foulant of low-pressure membranes. Moreover, the significant organic fouling that is contributed by polysaccharides and/or proteins in macromolecular and/or colloidal forms depends on molecular shape (structure) as well as size (i.e. molecular weight).

More significant flux decline was observed in microfiltration (MF) compared to ultrafiltration (UF) membrane filtration. MF membrane fouling may be caused by pore blockage associated with large (macromolecular) hydrophilic molecules and/or organic colloids.

In the case of UF membranes, the flux decline may be caused by sequential or simultaneous processes of surface (gel layer) coverage during filtration. Morphological analyses support the notion that membrane roughness may be considered as a more important factor in membrane fouling by controlling interaction between molecules and the membrane surface, compared to the hydrophobic/hydrophilic character of membranes. Membrane fouling mechanisms are not only a function of membrane type (MF versus UF) but also depend on source (feed) water characteristics.

Benchmarking of municipal waste water treatment plants (an Austrian project); Lindtner S et al.; An Austrian research project focused on the development of process indicators for treatment plants with different process and operation modes.

The whole treatment scheme was subdivided into four processes, i.e. mechanical pretreatment (Process 1), mechanical-biological waste water treatment (Process 2), sludge thickening and stabilisation (Process 3) and further sludge treatment and disposal (Process 4).

In order to get comparable process indicators it was necessary to subdivide the sample of 76 individual treatment plants all over Austria into five groups according to their mean organic load (COD) in the influent. The specific total yearly costs, the yearly operating costs and the yearly capital costs of the four processes have been related to the yearly average of the measured organic load expressed in COD (110 g COD/pe/d). The specific investment costs for the whole treatment plant and for Process 2 have been related to a calculated standard design capacity of the mechanical-biological part of the treatment plant expressed in COD.

The capital costs of processes 1, 3 and 4 have been related to the design capacity of the treatment plant. For each group (related to the size of the plant) a benchmark band has been defined for the total yearly costs, the total yearly operational costs and the total yearly capital costs. For the operational costs of the Processes 1 to 4 one benchmark ({see symbol in text} per pe/year) has been defined for each group. In addition a theoretical cost reduction potential has been calculated. The cost efficiency in regard to water protection and some special sub-processes such as aeration and sludge dewatering has been analysed.

A new side stream process for easily degradable industrial waste waters to avoid sludge bulking; Wandl G et al.; A new treatment scheme for the treatment of easily biodegradable industrial waste waters has been developed. The side stream treatment of dairy waste water with the excess sludge from the domestic treatment line of the regional treatment plant Bad Voslau has been operated successfully for a period of three years during which the industrial load stemming from the dairy increased from 800 kg COD/d to 2,500 kg COD/d with peak loads up to 5,000 kg/d.

Despite of the increased load to the treatment plant the

total aeration tank volume had not been increased. This treatment is performed in an existing aeration tank of the WWTP (V = 1,800 m3) which is now used as contact tank for the combined aeration of dairy waste water and excess sludge from the domestic treatment line (volume aeration tank = 15,000 m3). In this tank the easily degradable substrate from the industrial waste is mainly adsorbed to the biological sludge and after a mechanical dewatering transferred to the anaerobic digester where it yields in an increased gas production.

The filtrate of the dewatering process is completely free from biodegradable material and can without danger of bulking be fed to the aeration tank of the domestic treatment line. The new process has proven to be extremely flexible since already now daily peak loads exceeding the design load by more then 60% could be treated in the plant without any problems. Compared to other alternatives for the dairy waste water treatment that were investigated during this study, the new side stream process is very advantageous. No other pre-treatment process for industrial waste water could have been operated under comparable loading conditions without severe operating problems.

Degradation of endocrine disrupting chemicals bisphenol A, ethinyl estradiol, and estradiol during UV photolysis and advanced oxidation processes; Rosenfeldt EJ et al.; The degradation of three endocrine disrupting chemicals (EDCs), bisphenol A, ethinyl estradiol, and estradiol, was investigated via ultraviolet (UV) radiation photolysis and the UV/hydrogen peroxide advanced oxidation process (AOP).

These EDCs have been detected at low levels in wastewaters and surface waters in both the United States and European countries, can cause adverse effects on humans and wildlife via interactions with the endocrine system, and thus must be treated before entering the public drinking water supply. Because many EDCs can only be partially removed with conventional water treatment systems, there is a need to evaluate alternative treatment processes. For each EDC tested, direct UV photolysis quantum yields were derived for use with both monochromatic low-pressure (LP) UV lamps and polychromatic medium-pressure (MP) UV lamps and second-order hydroxyl radical rate constants were developed.

These parameters were utilized to successfully model UV treatment of the EDCs in Labouratory and natural waters. The polychromatic MP UV radiation source was more effective for direct photolysis degradation as compared to conventional LP UV lamps emitting monochromatic UV 254 nm radiation. However, in all cases the EDCs were more effectively degraded utilizing UV/H2O2 advanced oxidation as compared to direct UV photolysis treatment.

Accumulation of arsenic in drinking water distribution systems; Lytle DA et al.; The tendency for iron solid surfaces to adsorb arsenic is well-known and has become the basis for several drinking water treatment approaches that remove arsenic. It is reasonable to assume that iron-based solids, such as corrosion deposits present in drinking water distribution systems, have similar adsorptive properties and could therefore concentrate arsenic and potentially re-release it into the distribution system.

The arsenic composition of solids collected from drinking water distribution systems (pipe sections and hydrant flush solids), where the waters had measurable amounts of arsenic in their treated water, were determined. The elemental composition and mineralogy of 67 solid samples collected from 15 drinking water utilities located in Ohio (7), Michigan (7), and Indiana (1) were also determined. The arsenic content of these solids ranged from 10 to 13 650 microg of As/g of solid (as high as 1.37 wt %), and the major element of most solids was iron. Significant amounts of arsenic were even found in solids from systems that were exposed to relatively low concentrations of arsenic (<10 microg/L) in the water.

Determination of polar drug residues in sewage and surface water applying liquid chromatography-tandem mass spectrometry; Zuehlke S et al.; A simple and rapid method is presented for the trace-level analysis of 10 polar pharmaceutical residues in various types of water samples from the aquatic environment. Using this method, the pharmaceuticals and several drug metabolites can be Analysed in drinking and surface waters and in wastewater (treated and untreated sewage) at concentrations down to 0.01 microg/L.

Samples are prepared by a simple in situ derivatization enabling the preconcentration of very polar metabolites by automated solid-phase extraction. The analytes were separated by liquid chromatography with tandem mass spectrometric detection and quantified by comparison with an internal standard. Limits of quantification were between 0.01 and 0.02 microg/L for three phenazone-type pharmaceuticals, six of their metabolites, and the antiepileptic drug carbamazepine. Except for dimethylaminophenazone, recoveries for all analytes were between 87 and 117% for raw and purified sewage, groundwater, and surface and drinking water.

Investigations of some environmental samples revealed that sewage and surface water treatment causes a slight reduction of the concentrations of some analytes whereas other compounds were persistent during water treatment. Thus, some compounds were detected at the low-microgram per liter level in sewage effluents of wastewater treatment plants in Berlin (Germany) and were also found at high-nanogram per liter concentrations in Berlin surface water samples.

Rapid cyanide detection using the Cyantesmo kit; Rella J et al.; BACKGROUND: Sources of cyanide exposure are many, including combustion of plastic and vinyl, such as in a house fire, Labouratory or industrial exposures including exposure in the electroplating industry both of printed circuit boards and in jewelry work. Rapid and definitive diagnosis of cyanide poisoning is unavailable in the emergency department setting. It is desirable to make a definitive diagnosis in order to prevent potential complications of empiric treatment of presumptive cyanide poisoning from the cyanide antidote kit currently approved by the US Food and Drug Administration (FDA).

We investigated a technique to detect cyanide currently utilized by water treatment facilities to determine if it can be applied to rapidly detect concentrations of cyanide in the clinically important range. METHODS: Varying standardized dilutions of KCN ranging from 0.25 microg/mL to 30 microg/mL were acidified with a drop of sulphuric acid in a closed system under a ventilation hood. Cyantesmo test strips were placed into the test tubes above the fluid level where liberated HCN gas interacted with the test strip to effect a Colour change. Colour changes were compared to negative controls and to each other. RESULTS: The test strips demonstrated an incrementally increasing deep blue Colour change over a progressively longer portion of the test strip in less than 5 minutes for each concentration of KCN including 1, 3, 10, and 30 microg/mL.

The concentrations of 0.25, 0.5, and 0.75 required more than 2 hours to begin demonstration of any Colour change. CONCLUSION: The Cyantesmo test strips accurately and rapidly detected, in a semi-quantifiable manner, concentrations of CN greater than 1 microg/mL contained in each test sample. Future work to validate this test in blood and in clinical specimens is planned.

REPORTS

Investigation of the formation of chlorination by-products in water rich in bromide and organic matter content; Nikolaou AD; Bench-scale chlorination experiments with river water rich in bromide and organic matter content were performed in order to investigate the Behaviour and speciation of the chlorination by-products, especially the brominated ones, which are considered more harmful to human health than their chlorinated analogues.

The analysis of the compounds was performed by means of gas chromatography. Statistical treatment of the results with Multifactor Analysis of Variance technique revealed the trends of each compound as function of pH, chlorine dose, reaction time and temperature.

Brominated species of chlorination by-products predominated, due to the presence of bromide in the river water leading to preferable bromination than chlorination of the organic matter. Bromide incorporation factors were calculated for trihalomethanes and haloacetic acids, the major categories of compounds formed. While bromide incorporation factors decreased with increasing chlorine dose, the opposite was true for low range of chlorine dose, a fact that needs to be considered during treatment of waters with similar properties.

Bromide incorporation factors decreased with increasing pH, especially for haloacetic acids. While the organic matter content can be sufficiently removed during water treatment, bromide ion still remains in the treated water. The results of the present study indicate that chlorination conditions should be optimized in order to minimize the incorporation of bromide into the chlorination by-products formed, which would increase their toxic effects on the water consumers.

Environmental applications for biosurfactants; Mulligan CN; Biosurfactants are surfactants that are produced extracellularly or as part of the cell membrane by bacteria, yeasts and fungi. Examples include Pseudomonas aeruginosa which produces rhamnolipids, Candida (formerly Torulopsis) bombicola, one of the few yeasts to produce biosurfactants, which produces high yields of sophorolipids from vegetable oils and sugars and Bacillus subtilis which produces a lipopeptide called surfactin. This review includes environmental applications of these biosurfactants for soil and water treatment.

Biosurfactant applications in the environmental industries are promising due to their biodegradability, low toxicity and effectiveness in enhancing biodegradation and solubilization of low solubility compounds. However, more information is needed to be able to predict and model their behaviour. Full scale tests will be required. The role of biosurfactants in natural attenuation processes has not been determined. Very little information is available concerning the influence of soil components on the remediation process with biosurfactants. As most of the research until now has been performed with rhamnolipids, other biosurfactants need to be investigated as they may have more promising properties.

Equilibrium, kinetic and mass transfer studies and column operations for the removal of arsenic(III) from aqueous solutions using acid treated spent bleaching earth; Mahramanlioglu M et al.; In the present study, a new adsorbent was produced from spent bleaching earth by H2SO4 impregnation method. The sorption of arsenic(III) by acid treated spent bleaching earth was studied to examine the possibility of utilizing this material in water treatment systems.

The effect of time, pH, initial concentration, temperature on the adsorption of arsenic(III) was studied. Maximum adsorption was found to occur at pH 9.0. The adsorption process followed the first order Lagergren equation. Mass transfer coefficients and rate constants of intraparticle diffusion were calculated. The experimental data points were fitted to the Langmuir equation in order to calculate the adsorption capacity (Q0) of the adsorbent and the value of Q0 was found to be 0.46 mmol g(–1). In order to understand the adsorption mechanism, Dubinin-Radushkevich (DR) isotherm was used. The magnitude of E calculated from DR equation was found to be 5.12 kJ mol(–1).

The heat of adsorption (deltaH0 = –30367 J mol(–1)) implied that the adsorption was physical exothermic adsorption. The column studies were also carried out to simulate water treatment processes. The capacity values obtained in column studies were found to be greater than the capacity values obtained in batch studies. This result was explained by the difference between batch system and column system. The factors that affect the capacity values of column and batch systems were explained. The effect of other anions on the adsorption of arsenic(III) in the presence of NO3-, SO4(2-), Cl-, Br- was studied. The presence of these anions did not affect the adsorption of arsenic(III) significantly.

WASTE WATER

Wastewater sludge convective drying: influence of sludge origin; Leonard A et al.; This paper deals with thermal drying of wastewater sludges, whose management will become crucial in the forthcoming years. Sludges collected after mechanical dewatering in 5 different WWTPs are submitted to the same convective drying treatment in order to try finding some relations between the drying behaviours, the type of effluent and wastewater treatments. Results show dearly sludge drying remains a particularly complex operation because sludges may exhibit very different behaviours, both from kinetic and texture points of view. Initial moisture content and global composition are not sufficient to explain the different observed behaviours.

It is too early to claim one or another type of water treatment has an influence on the drying behaviour. However, the drying rates can be classified in the inverse order of the organic content. Moreover for sludges at the same siccities, the harder the material (rheological properties), the higher the drying rate. Final textural properties can be related to the rheological properties and the internal diffusion limitations lowering the drying intensity.

BY RADIATIONS

UV radiation-induced enterobacterial responses, other processes that influence UV tolerance and likely environmental significance; Rowbury RJ; The ability of enterobacteria to become UV-tolerant is important because such tolerance may enable organisms to resist irradiation in the environment, in water treatment, in shell-fish, in stages of food processing, and at locations in the domestic, commercial and hospital environment The mechanism for regulation of tolerance induction and SOS response induction has been studied for many years, and is well understood, except for the early stages of induction.

Such early stages, namely sensing of the stimulus (UV irradiation) and the way in which such sensing leads to signal production, have until now been poorly understood. The claim has been made that DNA is the sensor and that either damage to DNA or production of SS regions in DNA (following interaction of UV with DNA) triggers the signal that sets in train RecA activation and other stages of tolerance induction. This claimed induction mechanism is a "classical" one in the sense that it involves intracellular sensing (by DNA) of the stressing stimulus (UV), and production of an intracellular signalling molecule.

It is not, however, firmly established as the mechanism

for initiation of UV tolerance induction and SOS response induction. The results reviewed here give firm evidence for a different and unique mechanism for sensing of UV and production of the signal. These results establish without doubt that, for UV tolerance induction, the UV sensor is an extracellular protein, which is a UV tolerance-specific extracellular sensing component (ESC). This component is formed by unstressed cells and on interacting with the stimulus (UV) in the medium, is converted to the tolerance induction signalling molecule, which is a UV tolerance-specific extracellular induction component (EIC).

It is this extracellular signal which interacts with the sensitive organisms and triggers tolerance induction. This pair of extracellular components (ECs) may offer the only means of switching-on such tolerance induction; certainly they offer the only known way for early warning to be given of impending UV challenge. Thus, the EIC can diffuse from a region of UV stress to a stress-free region and there warn organisms of impending stress and prepare them to resist it. d, h, k, k. As indicated here, UV irradiation not only induces UV tolerance, but also switches-on acid tolerance, alkali tolerance and thermotolerance responses. The fact that all three responses involve ESC/EIC pairs strongly supports the view that functioning of such EC pairs form the major, if not the only, means for UV tolerance induction.

The UV tolerance-specific ESC can detect other stresses and becomes activated, leading to cross-tolerance responses. Of particular interest, this ESC acts as a biological thermometer, detecting increases in temperature, such increases leading to gradually increasing formation of the EIC and, accordingly, gradual increases in UV tolerance.

This UV tolerance-specific ESC can also detect other stresses e.g. acting as a pH sensor. In all cases, on activation, the EIC formed (from this specific ESC) only induces UV tolerance.

It is proposed that the interaction of EICs with stress-sensitive organisms should be examined, and it is suggested that such EICs may, directly or indirectly, interact with and activate the same stress response regulators as are used to detect internal stressors and which, on activation, also trigger the switching-on of stress responses. For example, EICs either a in a protonated or oxidised state (formed by activation of ESCs by H+ or H2O2) or b produced by irradiation, may lead to protonation or oxidation or other forms of activation of the appropriate regulator (e.g. Fur or OxyR or RecA etc), leading to response induction.

Bromate Control

Enhanced bromate control during ozonation: the chlorine-ammonia process; Buffle MO et al.; Potentially carcinogenic bromate forms during the ozonation of bromide-containing waters. Some water treatment facilities have had to use ammonia addition and pH depression to minimize bromate formation, but these processes may prove to be insufficient to comply with upcoming regulations. The chlorine-ammonia process (Cl2-NH3), consisting of prechlorination followed by ammonia addition priorto ozonation is shown to cause a 4-fold decrease in bromate formed when compared to the ammonia-only process.

Experiments revealed three key mechanisms:

- Oxidation by HOCl of Br- to HOBr and its subsequent masking by NH3 as NH2Br.
- Decrease of HO- exposure through halogenation of Dissolved Natural Organic Matter (DNOM) by HOCI and scavenging of HO by NH_2Cl.
- DNOM acting as a bromine sink after oxidation of Br- to HOBr.

At an ozone exposure of 6 mg/L × min and pH 8, conventional ozonation of Lake Zurich water spiked with 560 microg/L Br- formed 35 microg/L BrO3-, whereas the application of the Cl2-NH3 process resulted in 5 microg/L BrO3-. Additional pH depression to pH 6 further decreased bromate formation by a factor of 4. Trihalomethanes (THM) and cyanogen chloride (CNCl), that mayform during prechlorination and monochloramination, respectively, were well below regulatory limits. The chlorine-ammonia process holds strong promise for water treatment facilities struggling with a bromate formation problem during ozonation.

Assessment

Labouratory assessment of the mobility of nanomaterials in porous media; Lecoanet HF et al.; The production of significant quantities of engineered nanomaterials will inevitably result in the introduction of these materials to the environment. Mobility in a well-defined porous medium was evaluated for eight particulate products of nanochemistry to assess their potential for migration in porous media such as groundwater aquifers and water treatment plant filters.

Contrary to the assertion that nanomaterials present monolithic environmental risks, here we show that these nanomaterials exhibit widely differing transport Behaviours. Fullerene-based nanomaterials that had been functionalized to facilitate dispersal in water displayed the highest mobilities, with a calculated potential to migrate approximately 10 m in unfractured sand aquifers. Colloidal aggregates of C60, which have been the focus of recent toxicity studies, were among the least mobile of the nanomaterials evaluated.

Chloride effect on TNT degradation by zerovalent iron or zinc during water treatment; Hernandez R et al.; Addition of corrosion promoters, such as sodium and potassium chloride, accelerated TNT degradation during water treatment using zerovalent zinc and iron. It was theorized that corrosion promoters could be used to accelerate electron generation from metallic species, create new reactive sites on the surface of metals during contaminated water treatment, and minimize passivating effects.

The surface area normalized pseudo-first-order rate constant for the reaction of zerovalent zinc with TNT in the absence of KCl was 1.364 L × m(−2) × h(−1). In the presence of 0.3 mM and 3 mM KCI, the rate constant increased to 10.5 L × m(−2) × h(−1) and 51.0 L × m(−2) × h(−1), respectively. For the reaction with zerovalent iron and TNT, the rate constant increased from 6.5 (L/m2 x h) in the absence of KCl to 37 L × m(−2) × h(−1) using 3 mM KCl.

The results demonstrate that chloride based corrosion promoters enhance the rate of TNT degradation. The in-situ breakage of the oxide layer using corrosion promoters was applied as a treatment to maintain the long-term activity of the metallic species. Zinc maintained a high reactivity toward TNT, and the reactivity of iron increased after 5 treatment cycles using 3 mM KCI. Zinc and iron scanning electron micrographs indicate that TNT degradation rate enhancement is caused by the pitting corrosion mechanism.

Effectiveness

Determining the effectiveness of conventional and alternative coagulants through effective characterization schemes; Sinha S et al.; Polymeric forms of metal coagulants in water treatment have become increasingly used due to their wider availability and reduction in cost. These specialized coagulant forms and products are claimed by manufacturers to be superior to conventional coagulants in particulate and/ or organic removal with inherent advantages of lower alkalinity consumption and lesser sludge production. However, due to their proprietary nature, little is known about their chemical composition.

To determine and understand the effectiveness of these alternative coagulants, a comprehensive study was undertaken to characterize metal coagulants, and to comparatively evaluate them on a well-characterized source water. The objective of this study was to provide a scheme for utilities that could be employed as a screening process and a method of selecting an appropriate coagulant based on raw water characteristics and insight into the coagulatability of the source water.

Characterizations of coagulants included:

- Active metal content.
- Anion content.
- Acidity
- Alkalinity consumption.
- Charge reversal by colloidal titration.
- Molecular weight determination.

A total of five poly-aluminum chlorides (PACl), along with a conventional coagulant (aluminum sulfate or alum) were evaluated. Results show that through the characterization scheme, an effective coagulant (conventional versus alternative) and coagulant type (among various PACl) can be chosen before undertaking time-consuming bench or pilot-scale evaluation.

Chapter 6

Effect of Biologicals

BACKGROUND

Survival and injury of Arcobacter after artificial inoculation into drinking water; Moreno Y et al.; The aim of this work was to assess the effect of chlorine water treatment on Arcobacter butzleri and to study the survival strategies of this organism in chlorinated and non-chlorinated drinking water. A. butzleri NCTC 12481 was inoculated into chlorinated and non-chlorinated water and samples were removed aseptically, immediately and periodically during the next 2 days (for chlorinated drinking water) or 35 days (for non-chlorinated drinking water). The membrane integrity (Live/Dead BacLight kit), 16S rRNA (FISH technique), DNA content (23S rRNA PCR-RFLPs) and culturability changes in A. butzleri cells were Analysed. Culturability of the cells was lost at 5 min in chlorinated drinking water.

At that time the cells showed membrane damage, although fluorescent intensity of 16S rRNA hybridization was constant throughout the chlorine treatment. After 48 h the amplicon specific for the 23S rRNA gene was weakly detected. In non-chlorinated drinking water cells lost their culturability after 16 days but the other factors measured indicated that Arcobacter remained viable throughout the experiment.

INTRODUCTION

Environmental Protection Agency, and the Council of

State and Territorial Epidemiologists have maintained a colLabourative surveillance system for collecting and periodically reporting data related to occurrences and causes of waterborne-disease outbreaks (WBDOs) related to drinking water; tabulation of recreational water-associated outbreaks was added to the surveillance system in 1978.

This surveillance system is the primary source of data concerning the scope and effects of waterborne disease outbreaks on persons in the United States. This includes data on WBDOs associated with recreational water that occurred during January 2001-December 2002 and on a previously unreported outbreak that occurred during 1998. DESCRIPTION OF SYSTEM: Public health departments in the states, territories, localities, and the Freely Associated States are primarily responsible for detecting and investigating WBDOs and voluntarily reporting them to CDC on a standard form.

The surveillance system includes data for outbreaks associated with both drinking water and recreational water; only outbreaks associated with recreational water are reported in this summary.

During 2001-2002, a total of 65 WBDOs associated with recreational water were reported by 23 states. These 65 outbreaks caused illness among an estimated 2,536 persons; 61 persons were hospitalized, eight of whom died. This is the largest number of recreational water-associated outbreaks to occur since reporting began in 1978; the number of recreational water-associated outbreaks has increased significantly during this period ($p<0.01$). Of these 65 outbreaks, 30 (46.2%) involved gastroenteritis.

The etiologic agent was identified in 23 (76.7%) of these 30 outbreaks; 18 (60.0%) of the 30 were associated with swimming or wading pools. Eight (12.3%) of the 65 recreational water-associated disease outbreaks were attributed to single cases of primary amebic meningoencephalitis caused by Naegleria fowleri; all eight cases were fatal and were associated with swimming in a lake (n = seven; 87.5%) or river (n = one; 12.5%). Of the 65 outbreaks, 21 (32.3%) involved dermatitis; 20 (95.2%) of these 21 outbreaks were associated with spas or pools.

In addition, one outbreak of Pontiac fever associated with a spa was reported to CDC. Four (6.1%) of the 65 outbreaks involved acute respiratory illness associated with chemical exposure at pools. The 30 outbreaks involving gastroenteritis comprised the largest proportion of recreational water-associated outbreaks during this reporting period.

These outbreaks were associated most frequently with Cryptosporidium (50.0%) in treated water venues and with toxigenic Escherichia coli (25.0%) and norovirus (25.0%) in freshwater venues.

The increase in the number of outbreaks since 1993 could reflect improved surveillance and reporting at the local and state level, a true increase in the number of WBDOs, or a combination of these factors. PUBLIC HEALTH ACTION: CDC uses surveillance data to identify the etiologic agents, types of aquatics venues, water-treatment systems, and deficiencies associated with outbreaks and to evaluate the adequacy of efforts (e.g., regulations and public awareness activities) for providing safe recreational water.

Surveillance data are also used to establish public health prevention priorities, which might lead to improved water-quality regulations at the local, state, and federal levels.

CAUSE AND EFFECT

Re-evaluation and re-construction of water purification system using soil. II. Removal of pollutants from infiltrating water; Hamasaki T et al.; Presented in this paper are the results of a series of column experiments and a pilot scale test conducted to develop the treatment system of freshwater and wastewater by their percolation through soil and the other materials at the rate of a few metres per day. Materials that had high ability to sorb fulvic acid (FA hereafter) and phosphate ion were selected based on the results of the batch sorption tests and were processed into beads before subjecting them to the column test. The results of the column experiment suggested that the reagents added to increase the strength of the beads reduced the sorption of FA, thus causing the early breakthrough of the FA injected into the column.

Less effect from the bead processing was observed for the removal of phosphate ion. Pilot scale test showed that the removal of particulate fraction of phosphorus differed among the materials used. The bead processing technique that can ensure the high water permeability while maintaining the sorption capacity of the material was required for further development of the treatment system.

Re-evaluation and reconstruction of water purification system using soil. I. Assessment of soil as a sorbent of humic substances and phosphate ion; Fujikawa Y et al.; The purpose of our study is to develop a treatment procedure for humic substances (HS hereafter) and phosphate ion in wastewater and environmental water by percolation of the water through a constructed soil layer at the hydraulic loading of a few metres per day.

In the present work, batch sorption tests were conducted for more than 80 samples of soil, sludge, mineral and organic materials in order to find good sorbents for fulvic acid (FA hereafter) and phosphate ion. The results showed that the sorption of FA was high for some charcoal, and apatite and goethite minerals.

Comparatively high sorption of FA was found for some Andosols and volcanic ash soil. Significant sorption of phosphate ion, on the other hand, was found for various types of soil, sludge from water treatment plants and some waste materials. The linear isotherm was obtained for the sorption of FA to a charcoal, apatite and goethite minerals, and Andosols.

INTEGRATED SYSTEM

Integrated systems analysis of persistent polar pollutants in the water cycle; van der Voet E et al.; Persistent polar pollutants (P3) are difficult to degrade in standard waste water treatment plants. As a result, they end up in the effluent and are emitted to the surface water. In some areas, this problem is aggravated through "closed loop recycling", causing concentrations of P3 in surface water to build up over time.

This could cause violation of (future) EU regulations. In the P-THREE project, various alternative waste water treatment techniques are investigated regarding their effectiveness in eliminating these substances, especially membrane bioreactor treatment and advanced oxidation processes, MBR and AOP.

The integrated systems analysis which is the subject of this paper assesses these techniques in a broader systems context:

- The life-cycle of the P3.
- The life cycle of the WWTPs.
- The WWTP life cycle costs.

MASS SPECTROMETER

Mass spectrometric analytical methods for the determination of endocrine disrupting chemicals (EDCs); Stehmann A et al.; This paper describes a method for the determination of endocrine disrupting chemicals (EDCs) in waste water. The method involves a SPE-C18 extraction followed by a SPE-silica gel column clean up and elution of the analytes with a mixture of acetone/pentane (2+1).

Thereafter derivatisation of the analytes with heptafluorobutyric acid anhydride (HFBA) and analysis by gas chromatography/mass spectrometry (GC/MS) with positive electron impact ionisation (EI) or negative chemical ionisation (NCI) using methane as reactant gas was performed. The two different ionisation techniques were compared and the negative chemical ionisation technique proved significant lower limits of detection (LODs) and quantitation (LOQs) than positive electron impact ionisation.

Recoveries of the analytes in various spiked waste water samples ranged from 75.4 to 96.9%. The concentrations of the EDCs in feed waste water from three different waste water treatment plants (WWTPs) ranged from non-detectable up to a concentration of 10,305 ng/L for 4-NP and in the corresponding effluent samples from non-detectable to 723 ng/L for BPA. Behaviour of bisphenol A (BPA), 4-nonylphenol (4-NP) and 4-nonylphenol ethoxylates (4-NP1EO, 4-NP2EO) in oxidative water treatment processes; Lenz K et al.; Endocrine disrupting chemicals (EDCs) such as natural and synthetic hormones or industrial chemicals can adversely affect the endocrine system at very low concentrations. As such substances can be present in raw water used for drinking-water production, they potentially pose a health risk to humans.

In this study Labouratory tests were performed to determine removal efficiencies of selected oxidative drinking water treatment processes, namely ozonation (1.4 mg/l O3) and chlorination, using sodium hypochlorite (0.5 mg/l NaClO) and chlorine dioxide (0.4-0.6 mg/l ClO2) under conditions applied in technical plants. 500-300,000 ng/l of bisphenol A (BPA), 4-nonylphenol (4-NP) and 4-nonylphenol-n-ethoxylates (NPnEO) were selected for investigations and measured by HPLC/FLD and HPLC/MS.

To investigate possible oxidation by-products, adsorbable organic halogens (AOX) were determined and estrogenic activities were assessed with the help of an estrogen receptor binding assay (YES). Ozonation and chlorination with ClO2 removed both 4-NP and BPA below detection limits, corresponding with AOX and estrogenic activity.

Concerning NPnEO ozonation removed NP1EO and NP2EO up to 28% and 30%, respectively, whereas ClO2 showed high removal efficiencies, eliminating >94% and 92%, respectively. NaCIO removed 4-NP and BPA below detection limits, but estrogenic activities increased and AOX could be measured. NP1EO and NP2EO were only marginally reduced corresponding to the slight decrease of estrogenic potential.

VACCINES

Vaccines against diarrheal diseases; Nataro JP; At least 2 million persons succumb annually to enteric infection, and in countless other patients, diarrheal disease aggravates malnutrition and susceptibility to other infections. Prevention of enteric illness by virtue of improved hygiene and provision of sanitation and water treatment is impractical in most developing countries, where morbidity and mortality rates are highest. For this reason, development of vaccines against the most important gastrointestinal infections remains a high priority.

Recent advances in removing phosphorus from wastewater and its future use as fertilizer (1997-2003); de-Bashan LE et al.; Large quantities of phosphate present in wastewater is one of the main causes of eutrophication that negatively affects many natural water bodies, both fresh water and marine.

It is desirable that water treatment facilities remove phosphorus from the wastewater before they are returned to the environment. Total removal or at least a significant reduction of phosphorus is obligatory, if not always fulfilled, in most countries. This comprehensive review summarizes the current status in phosphorus-removal technologies from the most common approaches, like metal precipitation, constructed wetland systems, adsorption by various microorganisms either in a free state or immobilized in polysaccharide gels, to enhanced biological phosphorus removal using activated sludge systems, and several innovative engineering solutions.

As chemical precipitation renders the precipitates difficult, if not impossible, to recycle in an economical industrial manner, biological removal opens opportunities for recovering most of the phosphorus and beneficial applications of the product. This review includes the options of struvite (ammonium-magnesium-phosphate) and hydroxyapatite formation and other feasible options using, the now largely regarded contaminant, phosphorus in wastewater, as a raw material for the fertilizer industry. Besides updating our knowledge, this review critically evaluates the advantage and difficulties behind each treatment and indicates some of the most relevant open questions for future research.

Chlorination remains the most common form of wastewater disinfection in North America due to its low cost and long-term history of effectiveness. One disadvantage is that chlorination of residual organic material can generate chlorinated-organic compounds that may be carcinogenic or harmful to the environment. Residual chlorine or chloramines may also be capable of chlorinating organic material in the natural aquatic environment. Further, because residual chlorine is toxic to aquatic species, the treated effluent must also be chemically dechlorinated, adding to the complexity and cost of treatment.

Ultraviolet Light is becoming the most common means of disinfection in the UK because of the concerns about the impacts of chlorine in chlorinating residual organics in the wastewater and in chlorinating organics in the receiving water. Ultraviolet radiation is used to damage the genetic structure of bacteria, viruses, and parasites, making them incapable of reproducing.

The key disadvantages of ultraviolet disinfection is the need for frequent lamp maintenance and replacement and the need for a highly treated effluent to ensure that the ultraviolet radiation can get through to any microorganisms present (i.e. solids present in the treated effluent may protect microorganisms from the U.V. radiation). Ozone O_3 is generated by passing oxygen O_2 through a high voltage potential resulting in a third oxygen atom becoming attached and forming O_3. Ozone is very unstable and reactive and oxidizes most organic material it comes in contact with, destroying many disease-causing microorganisms.

It has the added bonus of removing other wastewater components such as colour. Ozone is considered to be safer than chlorine because, unlike chlorine which has to be stored on site, ozone is generated as it is required. Sludge treatment The coarse primary solids and secondary biosolids (bacteria) accumulated in a wastewater treatment process must be treated and disposed of in a safe and effective manner. This material is often inadvertently contaminated with toxic organic and inorganic compounds (e.g. heavy metals). The purpose of digestion is to reduce the amount of organic matter and the number of disease-causing microorganisms present in the solids. The most common treatment options include anaerobic digestion, aerobic digestion, and composting.

Anaerobic digestion Anaerobic digestion is a bacterial process that is carried out in the absence of oxygen. The process can either be thermophilic digestion (in which sludge is fermented in tanks heated to about 380°C) or mesophilic digestion (cold digestion of sludge where sludge is maintained in large tanks for weeks to allow natural mineralisation of the sludge).

Thermophilic digestion generates biogas with a high proportion of methane that may be used to both heat the tank and run engines for other on-site processes. In large treatment plants sufficient energy can be generated in this way to produce electricity for sale. The methane generation is a key advantage of the anaerobic process. Its key disadvantage is the long time required for the process (up to 30 days) and the high capital cost.

Aerobic digestion Aerobic digestion is a bacterial process that runs in the presence of oxygen. Under aerobic conditions, bacteria rapidly consume organic matter and convert it into carbon dioxide. Because the aerobic digestion occurs much faster than anaerobic digestion, the capital costs of aerobic digestion are lower. However, the operating costs are characteristically much greater for aerobic digestion because of the need to add oxygen to the process.

Composting Composting is also an aerobic process that involves mixing the wastewater solids with sources of carbon such as sawdust or wood chips. In the presence of oxygen, bacteria digest both the wastewater solids and the added carbon source and, in doing so, produce a large amount of heat. Properly designed and controlled, the heat generated can be sufficient to significantly destroy a sufficient number of the disease-causing microorganisms to enable the resulting composted product to be safely used as a soil amendment material (with similar benefits to peat) for agricultural use.

Both anaerobic and aerobic digestion processes can result in the destruction of disease-causing microorganisms to a sufficient level to allow the resulting digested solids to be safely applied to land or used for agriculture as a fertilizer provided that levels of toxic constituents are sufficiently low.

The choice of a wastewater solid treatment method depends on the amount of solids generated and other site-specific conditions. However, in general, composting is most often applied to smaller-scale applications followed by aerobic digestion and then lastly anaerobic digestion for the larger-scale municipal applications.

Sludge Disposal When a liquid sludge is produced, further treatment may be required to make it suitable for final disposal. This includes lagooning in drying beds to produce a cake that can be applied to land or incinerated; pressing, where sludge is mechanically filtered, often through cloth screens to produce a firm cake; or liquid injection to land or liquid disposal to landfill. There are

concerns about sludge Incineration because of the contents of the emmissions to air, along with the high cost of supplemental fuel, making this a less attractive and less commonly constructed means of sludge treatment and disposal.

TREATMENT

Treatment in the receiving Environment Many processes in a wastewater treatment plant are designed to mimic the natural treatment processes that occur in the environment, whether that environment is a natural water body or the ground.

If not overloaded, bacteria in the environment will consume organic contaminants, although this will reduce the levels of oxygen in the water and may significantly change the overall ecology of the receiving water. Native bacteria feed on the organic contaminants, and disease-causing microorganisms are reduced by natural environmental conditions that are hostile to these organisms (microbial predation, ultraviolet radiation, etc.)

Consequently in cases where the receiving environment provides a high level of dilution, a high degree of wastewater treatment is not necessarily required. However, recent evidence has demonstrated that very low levels of certain contaminants in waste water, including hormones (especially from animal husbandry) and synthetic materials such as pthalates that mimic hormones in their action, can have an unpredictable adverse impact on the natural biota and potentially on humans if the water is re-used for drinking water.

However, in nature, microbes live in complex structured communities of diverse, metabolically redundant interacting species, mostly growing as biofilms on surfaces in nutrient-poor, heterogeneous matrices. Most microbes cannot yet be cultivated in the Labouratory. The challenge, then, is to exploit cultivation-independent methods to gain a mechanistic understanding of functional relationships in microbial biofilm communities in natural or near-natural settings, to elucidate the rules of microbial behaviour and interactions, and to understand the functional/evolutionary basis of microbial diversity. Genomics has proven to be a powerful, cultivation-independent approach for the characterization of cellular functions and metabolic potential.

Index